FULLERENE C_{60}

HISTORY, PHYSICS, NANOBIOLOGY, NANOTECHNOLOGY

FULLERENE C_{60}

HISTORY, PHYSICS, NANOBIOLOGY, NANOTECHNOLOGY

DJURO KORUGA

University of Arizona
Tucson, Arizona
U.S.A.

STUART HAMEROFF

University of Arizona
Tucson, Arizona
U.S.A.

JAMES WITHERS

MER Corporation
Tucson, Arizona
U.S.A.

RAOULF LOUTFY

MER Corporation
Tucson, Arizona
U.S.A.

MALUR SUNDARESHAN

University of Arizona
Tucson, Arizona
U.S.A.

1993

NORTH-HOLLAND

AMSTERDAM · LONDON · NEW YORK · TOKYO

ELSEVIER SCIENCE PUBLISHERS B.V.
Sara Burgerhartstraat 25
P.O. Box 211, 1000 AE Amsterdam, The Netherlands

Front cover illustration: © Rick Alan, 1993

ISBN: 0 444 89833 6

This book is printed on acid-free paper.

Printed in The Netherlands.

C_{60}, twinkle star!
Tell me now who you are.

I am smiling at the Sun,
I love thee;

I am goddess of the New Moon,
I love thee;

I am a calm lightning point:
The Golden Mean!

As a Being of the Sun,
That you have seen.

PREFACE

Science and technology related to Fullerene C_{60} and related compounds have attracted the attention and research interest of hundreds, perhaps thousands, of scientists and engineers over the past several years. This attraction stems from the Fullerenes' beautiful symmetry, dazzling chemical and physical properties, and potential utilization in nanotechnology. For example, Fullerene devices may emulate nanoscale behaviors of living systems! The authors of this book are university professors with diverse backgrounds (Koruga, Hameroff and Sundareshan, with primary research interests in the fields of C_{60}-nanotechnology, molecular computing and control engineering, respectively) and experts in production and technology of Fullerenes (Loutfy and Withers, with 17 and 25 US patents in the field of high technologies).

Although Buckminster Fuller, from whom Fullerenes get their name, lived in this century (his life is reviewed in Chapter 1), the Fullerene story (with its ties to symmetry) predates modern science. The history of fivefold symmetry and the icosahedron as mathematical-philosophical concept started with the ancient civilizations of Egypt, China, India, Greece, and the Mayans. Ancient Egyptian civilization had both spatial icosahedrons and temporal icosahedrons which they presented in space as orthogonal, (the great pyramids are one example of this geometry). Modern history of fivefold symmetry and the icosahedron began between 1984 and 1985, when Shechtman and his research team opened a new branch in crystallography (fivefold symmetry) and when the Kroto/Smalley research team discovered the C_{60} molecule (truncated icosahedron). The production of solid C_{60} by the Huffman/Krätschmer research team in 1990 provided a new stimulus for research by producing C_{60} in macroscopic amounts for use by the scientific and technological community. This achievement led to developments such as Koruga's August 1992 creation of the dimer C_{116} using scanning tunnelling engineering and Loutfy's hydrogenation of C_{60} and construction of the first Ni/C_{60} rechargeable batteries in December, 1992. New inventions based on C_{60} will continue to be forthcoming, particularly in the areas of superconductivity, quantum devices, and molecular electronic devices. The discovery of the C_{60} molecule (Kroto/Smalley), production of solid C_{60} (Huffman/Krätschmer) and technological inventions such as C_{116} (Koruga) have been serendipitous creations. A short history of "C_{60} serendipity" is detailed in this book.

Also included are the results of the authors' Fullerene research efforts, including atomic resolution images of Fullerene C_{60}, production of nanotubes, hydrogenation of C_{60}, Ni/C_{60} batteries, nanotechnology of C_{60}, comparison of C_{60} with biological systems, and others. Fullerene C_{60} is linked to nanobiology for two reasons: (1) Certain intracellular proteins called clathrin have the same symmetry (truncated icosahedron) as the C_{60} molecule. Within neuronal synapses, clathrin interacts with microtubules which, from standpoints of symmetry and information are "icosahedron shadows." We term these nanobiological structures with symmetry and structural links to C_{60} "bio-Fullerenes." (2) We expect C_{60} and other Fullerenes to have applications in molecular information devices, and researchers in Fullerene science and technology should be aware of possible mechanisms of information processing in bio-Fullerenes. Fullerene C_{60} technology on a nanometer scale will require control engineering. For this reason, we include an overview of control systems--in particular, general and optimal control of the Schrödinger equation. In addition, some experimental and theoretical work of other researchers are also presented.

The book is organized into three parts. Part One consists of four chapters, and Parts Two and Three have three chapters each. **Part One**, The Science of Fullerene C_{60}, includes: (1) The History of Fullerene C_{60}; (2) The Icosahedron: Symmetry and Golden Mean Properties; (3) Symmetry of Electronic and Vibrational-Rotational Spectra of C_{60}; and (4) STM Imaging and Vibrational-Rotational Spectra of C_{60}. **Part Two**, From Nanobiology to Nanotechnology, contains (5) Bio-Fullerenes; (6) Intelligence Within the Cell: Bio-Fullerene Information Processing; and (7) From Nanobiology to Nanotechnology: Basic Concept. And **Part Three**, Fullerene C_{60}: Production, Technology, and Applications, covers (8) Fullerene Production, (9) Nanotechnology of C_{60}, and (10) Applications of Fullerenes.

The front cover illustration is based on art work by Rick Alan. The purple color represents the C_{60} molecule in solution (toluene), while the yellow represents C_{60} in the crystal state. The back cover illustration is art work by Yousry Saleh based on an ancient Egyptian picture in which some elements possess Golden Mean properties. The author of the C_{60} poem is Djuro Koruga.

Sources of figures which were taken from the literature are listed at the end of the book. References are listed alphabetically. We hope to have captured not only the scientific and technological implications of Fullerenes, but the aesthetics as well.

ACKNOWLEDGEMENTS

Without several key people, this book could not have been produced in its present quality and form. We would like to thank Jovana Simić-Krstić, Mirko Trifunović and Svetlana Janković, who stimulated discussion and participated in the STM research, Chuck Hassen, who helped write Chapter 10, John Nemmers, who participated in the mass spectroscopy research of Fullerenes, Michael Rush, our copy editor, who heroically produced coherent text, and Richard Hofstad, who prepared the camera ready copy.

We are grateful to Barbara Kosta and Karin Glinsky, who helped find the Göethe poems used in the book, Ljubomir Vukosavljević, who assisted in translation, and Dyan Louria, who stimulated discussion and helped prepare the glossary, index, and references.

We also wish to express appreciation for permission to reproduce previously published and/or copyrighted material from the journals and books which are listed in figure notations and at the end of the book.

We also thank Michiel Bom, our editor, and Elsevier for publishing this book. Finally, we heartfully appreciate our families for enduring hardships during the book's preparation.

Tucson, Arizona
January 25, 1993

Djuro Koruga
Stuart Hameroff
Jim Withers
Raoulf Loutfy
Malur Sundareshan

CONTENTS

PART I

SCIENCE OF FULLERENE C$_{60}$

Universe is not a system.
Universe is not a shape.
Universe is a scenario.
You are always in Universe.
You can only get out of systems.

– Buckminster Fuller

CHAPTER 1

HISTORY OF FULLERENE C_{60}

The words *mass, energy,* and *information* are well known in both the scientific and public communities, but another word, similar to energy, is less known. **Synergy** has been defined as the *behavior of integral, aggregate, whole systems unpredicted by behaviors of any of their components or subassemblies of their components taken separately from the whole* by **Buckminster Fuller**, who became famous in the public community after EXPO '67 in Montreal. He developed the Dymaxion map, which, unlike commonly known maps, is not a simple shadow projection but a **topological** transformation of the *mass* and *energy* of a body surface through equivalent entities on the faces of a polyhedron. In this kind of process exists an *invariant* which we can call **topological information**, which provides a link between **parts, process, whole** and a new relationship between **part-whole**. Fuller's name became even more famous in 1985 when Kroto and his coworkers named the C$_{60}$ molecule "Buckminsterfullerene." This book, **Fullerene C$_{60}$**: *History, Physics, Nanobiology, Nanotechnology,* is one of the first to be published in this exciting field; it includes a short history of scientific discovery, as well as one possible answer to the question: for what purposes can C$_{60}$ be utilized?

F:1-1

Buckminster Fuller -- Bucky in time went to Harvard

**Middlesex Football Team in which Bucky (lower left)
was an active player**

SOURCE: Alden Hatch, <u>Buckminster Fuller: At home in the Universe</u>, Crown Pub. Inc., New York, 1974.

1.1 Buckminster Fuller: Life and Synergy Concept

In 1973, Alden Hatch wrote a book entitled **Buckminster Fuller:** *At home in the Universe* about the man, his ideas and his inventions. The first page of the book is illustrated with an **icosahedron**. Fuller did serious research regarding the icosahedron in 1948. His depiction of the icosahedron as a Thirty-One Great-Circle Triangular Grid was drawn on May 1, 1948. Based on a vector equilibrium approach, he defined the icosahedron as 31 "great circle railroad tracks of energy." He proposed that perhaps "the 31 great circles of the icosahedron lock up the energy changes of the electron, while the six great circles release the sparks." He also speculated that "The icosahedron makes it possible to have individuality in Universe. The vector equilibrium never pauses at equilibrium, but our consciousness is caught in the icosahedron when mind closes the switch. The icosahedron's function in Universe may be to throw the switch of cosmic energy into a local shunting circuit. In the icosahedron, energy gets itself locked up even more by the six great circles-- which may explain why electrons are borrowable and independent of the proton-neutron group." Fuller's ideas on the icosahedron cannot be considered in detail in this book but perhaps will be in a future one.

In 1638, Tomas Fuller, a Lieutenant in the British Army, was the first of the Fuller family to come to Massachusetts. Richard Fuller was born on February 13, 1861. He was a charming gentleman who married Caroline Wolcott Andrews of Chicago in 1891, and four years later, on July 12, 1895, they heard the cry of their first son, Richard Buckminster Fuller, Jr. in Milton, Massachusetts, which today is a suburb of Boston. Very early on, his nickname became "Bucky." He was a very sensitive child.

The town of Milton had an excellent football team, and when he turned eighteen, Bucky, despite his constitution and small size, went out for it. He was stubbornly independent in his thinking. In school when his teacher made a cube on the blackboard, starting from a dot, which she said did not exist, he could not see how it was possible to make something real out of nothing. This way of thinking was a "twinkle" of his invention of his own synergetic geometry many years later. Bucky liked a lot of different sports, including track and field, football, and kayaking. Despite the fact that one of his legs was two centimeters shorter than the other, he became the quarterback of his varsity football team.

F:1-2

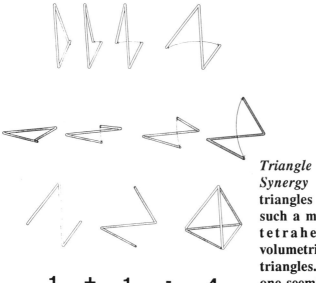

Triangle and Tetrahedron: Synergy (1+1=4): Two triangles may be combined in such a manner as to create the tetrahedron, a figure volumetrically embraced by four triangles. Therefore one plus one seemingly equals four.

$$1 \quad + \quad 1 \quad = \quad 4$$

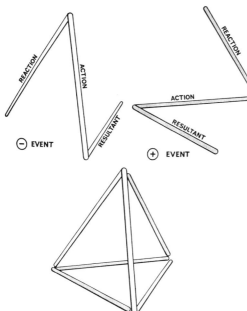

A TRIANGLE IS A SPIRAL
AND IS ONE ENERGY EVENT

ONE POSITIVE + ONE NEGATIVE EVENT
= TETRAHEDRON

Two Triangular Energy Events Make Tetrahedron: The open-ended triangular spiral can be considered one "energy event" consisting of an action, reaction and resultant. Two such events (one positive and one negative) combine to form the tetrahedron.

SOURCE: R. Buckminster Fuller, <u>Synergetics: Explorations in the Geometry of Thinking</u>, MacMillan Publishing Company, New York, 1975.

In 1913, he went to Harvard with a good scholastic and sports record. He was in the Navy during World War I, where his duties included serving as communications officer on the *George Washington*.   Afterwards, his first factory was the Red Rock Hall barn.   With much hard work, he invented machinery for mass production (the Stockade Building System), which was one of the first of some twenty-odd patents he was to acquire (U.S. Patent No. 1,633,702). With his first and second patents, he was ready for business; however, four years later, in 1927, he encountered his first financial difficulties, from which he was later able to recover.

From his engineering experience, Bucky had learned that the strength of various materials was amazingly greater in tension than in compression. He was fascinated with the design and engineering of Zeppelins. In 1928, he expressed ideas about "artificial life" as a 4D living machine, published as an article and drawings by the Chicago Evening Post. He received a great deal of publicity, and some financiers agreed to support his project. With their financial backing, he formed the "4D Company." Soon he had built a scale model of the 4D house; it was the first step towards his Dymaxion World.

Bucky accepted Einstein's 4D concept in physics and one of the possible solutions--the finite universe as a sphere. He thought strongly about Newton's and Einstein's universes and even wrote a poem about them:

Newton was a noun
And Einstein is a verb
Einstein's norm makes Newton's norm
Instant universe
Absurd
A body persists.
In a state of rest
Or--
Except as affected
Thus gravestones are erected!

Non-simultaneous, physical universe
Is energy, and
Energy equals mass
Times the second power

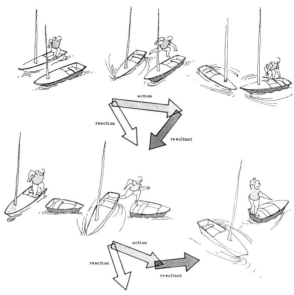

One Energy Event: Action, Reaction, and Resultant: One
energy event as demonstrated by the man jumping from the
boat. His action always demonstrates the action, reaction, and
resultant of the open-ended triangular spiral.

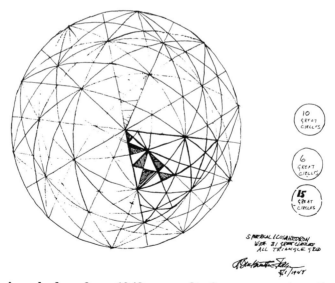

Fuller icosahedron from 1948 as result of energy event: action,
reaction, and resultant. He has descovered 31 great circles in
it.

SOURCE: R. Buckminster Fuller, <u>Synergetics: Explorations in the Geometry of Thinking</u>, MacMillan Publishing Company, New York, 1975.

of the speed of light
No exceptions!
Fission verified Einstein's hypothesis
Change is normal
Thank you, Albert!

After that he did research on sphere packing and discovered that if one removes the center sphere from a vector equilibrium, it closes down to an icosahedron as a twenty-sided "solid" structure. An icosahedron can be decomposed to an octahedron (an eight-sided figure) and this can be decomposed into tetrahedra. This means that all polyhedra can be subdivided into basic components--tetrahedra. "The tetrahedron is the basic structural system, and all structure in the universe is made up of tetrahedronal parts." He was delighted when he found a link between triangles and spheres: "When I had worked out my three-way great circle grid of perfect triangles for the map, I realized that it would be possible to make a spherical structure or dome, which is a half sphere, formed by thirty-one great circles around a sphere, crisscrossing each other to form perfect triangles of varying sizes. The surface triangles of a sphere are the outer parts of a tetrahedron. So I had all these tetrahedra and they all go to the center of the sphere; they cannot come apart."

From 1965 to 1967, Bucky and Shoji Sodao worked on the United States Pavilion at EXPO '67 in Montreal. They had the help of some young architects and engineers from MIT. A few years before, Bucky had met Shoji as a young Japanese student at the Cornell Architectural School. They soon became friends, and then partners (Fuller and Sodao, Inc., USA). Shoji was very modern in architectural ideas but, as is common with many Japanese, very traditional in his personal life.

Shoji Sadao was "a herald of spring" for Japanese scientific entry into the field of the icosahedron and C$_{60}$ molecule. As best we can determine today based on scientific publications, the first scientist to predict the existence of the C$_{60}$ molecule was Eiji Osawa.

1.2 Eiji Osawa

In April of 1971, Barth and Lawton published a paper in the *Journal of the American Chemical Society*, "The Synthesis of Corannulene," based on their previous work of 1966 and also on papers by Gleicher (1967) and Dewar and Gleicher (1966). **Corannulene** was of scientific interest because it contains a *pentagon* and possesses a unique electronic distribution.   It is a highly symmetrical molecule, with a non-planar configuration and nonalternate arrangement of sp^2 carbon atoms.

Based on these data and his own theoretical research, Eiji Osawa published two papers in Japanese about superaromatic hydrocarbons in 1970 and 1971. Karfunkel and Dressler (1992) correctly state that "Carbon does exist in several allotropic forms out of which C_{60} and C_{70} are the most recent representatives. C_{60} had been predicted by Osawa already in 1970. His paper, written in Japanese, attracted little attention due to language barriers and also due to the role of computational chemistry at that time: a quantitative and reliable confirmation of a reasonable suggestion by computational methods was intractable in 1970 (at least for molecules of that size)." In a 1970 paper, Osawa wrote about the *eicosahedron* and *truncated eicosahedron*, but he didn't write specifically about C_{60} because his main purpose was to explain superaromatic hydrocarbons. In 1971, Osawa (with Yoshida) published a book, *Aromaticity*, in which he considered the possibility of realizing overlap, "not between sigma bonds on the sphere, but between p_z orbitals directed perpendicular to surface of sphere." He proposed to truncate a Platonic icosahedron to produce a spherical structure with regular pentagons, and suggested the name *truncated icosahedrone* which "has the same design as that appearing on the surface of an official soccer ball." Bearing in mind that his main title was **"Aromaticity,"** he tried to provide an answer to the question: "How much aromaticity will be lost if the molecule is distorted from the strongly aromatic planar structure to a nonplanar structure?" He stressed that "the answer to this question must be a key to the possible existence of superaromaticity in C_{60} molecule." In the original (Japanese) version, he proposed the existence of the $C_{60}H_{60}$ molecule, which is very interesting for hydrogen storage and its application for Fullerene technologies, but in his English version, he mentions in a footnote that "$C_{60}H_{60}$ in the original print is the result of typographical error."

The idea of the icosahedron, pentagons, and sphere was born in the

F:1-4

The Synthesis of Corannulene
Wayne Barth and Richard Lawton

Barth WE and Lawton RG have synthesized molecule named corannulene, from the Latin: cor - heart and annula - rings. All the bond lengths, as a planar molecule, are equivalent 0.14 nm. The crucial points in this approach are: the central five-membered ring and "escape" from planar configurations of molecule to a bowl-shaped structure.

SOURCE: Barth WE and Lawton RG, The synthesis of Corannulene, J of the Am Chem Soc 93(7):1730, 1971.

FROM CORANNULENE TO SOCCER BALL
Eiji Osawa

2-3 corannulene

多数のベンゼン環の縮合した型のいわゆる "縮合多 B 式芳香族炭化水素" は典型的な平面分子である．これ らの代表的なベンゼノイド芳香族が球状分子の型をとった なら超芳香族性を示さないだろうか？ たとえばサッ カーの公式ボールの表面に描かれている幾何模様を思い 浮かべてみよう．それは正多面体として cube のつぎに小 さな正二十面体 (eicosahedron) (12)の頂点を全部切り 落として正五角形を出したもので，truncated eicosa hedron とでも称されるべき美しい多面体である(13)． 図ではわかりにくいところもあるので，もし手もとに サッカーボールがあれば手にとってながめていただくとは っきりするが，五角形(黒く塗ってある)の間には規則正 しく六角形がうずまっている．一見これらの成分多角形 はたいして曲がってもいないし，各辺はすべて同じ長さ にすることができる．もしこれらの頂点を全部 sp² 炭

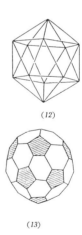

(12)

(13)

It is interesting how Osawa predicted C₆₀ existence as a truncated icosahedron. "Non-benzenoid aromatics were favourite targets for organic chemists in the 1960s and 1970s"--says Osawa. "My personal goal in those days was to find a new system more aromatic than the king benzene. Then came news of the corannulene synthesis by Lawton and the bowl-shaped structure of the molecule aroused strong interests in me: is it possible to extend this piece into a sphere so that one can realize a three-dimensional aromaticity? Luckily, my small son was playing with a football and I immediately recognized the pattern of the corannulene molecule in it. I started to study the whole design and soon learned that it is a truncated icosahedron . . . An interesting fact about this incident was that (for me, unfortunately) I discontinued the study of this molecule."

SOURCE: (1)   Osawa E, Kagaku (Kyoto) 25:854, 1970.

              (2)   Osawa E, The evolution of the football structure for the C₆₀ molecule: an historical perspective, The Royal Society Meeting
                      (London) on a Postbuckminsterfullerene view of the chemistry, Physics and Astrophysics of Carbon, October 1992.

"synergetic mind" of Buckminster Fuller. The idea then "travelled" from the American continent through Asia (Osawa - Japan) to Russia (Institute of Heteroorganic Compounds, Academy of Sciences in Moscow).

1.3 Bochvar and Gal'pern

These two Russian scientists submitted an original article for publication to Doklady Akademii Nauk SSSR in May 1972. The title of their paper was "Hypothetical Systems: Carbododecahedron, s-Icosahedron, and Carbo-s-Icosahedron," based on the experimental work of Eaton and Mueller (1972). These American scientists had pursued the synthesis of dodecahedrone and proposed the name *peristylane* for a group of columns arranged about an open space and designed to support a roof. They had synthesized a fragment of dodecahedrone containing six condensed five-membered rings. This experimental result and their knowledge of Archimodes' polyhedrons became the starting point for their research. They proposed the *truncated icosahedron*, which they called *s-icosahedron* (12 pentagons, 20 hexagons, 60 vertices and 90 edges). This molecule would be $C_{60}H_{60}$ and "carbon atoms in this system will be close to sp^2-hybridized." They defined ϱ-electrons, in carbo-s-icosahedron, as analogous to π-bonds in conjugated systems. They first calculated energy states of these molecules and concluded "that carbo-s-icosahedron should be more stable than carbododecahedron."

1.4 From Jones to Kroto

At the very time that Fuller was working on EXPO '67 for Montreal (1966), David Jones wrote an article in his ancestors' native country (England) under the pseudonym "Deadalus" in the journal *New Scientist* about the possible synthesis of spherical molecules, like little footballs, from graphite. His calculations indicated that the molecule should be 100 nm, which is a hundred times larger than C_{60}. His intuitive imagination (from graphite to spherical molecule like a soccer ball) surpassed his scientific results based upon calculation (size of spherical molecule). But as Einstein has noted, intuition is sometimes more important than knowledge for human progress, because knowledge is limited, while correct intuition is invariant.

F:1-5   **THE PERISTYLANE SYSTEM**
                              Philip Eaton and Richard Mueller

          1   2

"We are pursuing the synthesis of dodecahedrane (1). For the purpose of synthetic design we view dodecahedrane as made up of two subunits, cyclopentane and the C-15 fragment 2, as illustrated. We have named the larger fragment "peristylane" from the Greek περιςτυλου, a group of columns arranged about an open space and designed to support a roof."

SOURCE: Eaton PE and Mueller RH, The Peristylane system, J of the Am Chem Soc 94(3):1014, 1972.

FROM PERISTYLANE SYSTEM TO S-ICOSAHEDRON
D.A. Bochvar and E.G. Gal'pern

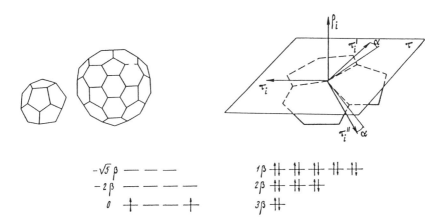

The first calculation of C$_{60}$ molecule was done by Bochvar and Gal'pern in 1972. They compared dodecahedron (1) and s-icosahedron (2) and showed disposition of orbitals of the i-th atom of s-icosahedrane (3). Based on the energy level diagram (4) they concluded "in contrast to the carbodecahedron calculations, calculations on the ʃ-system of carbo-s-icosahedron revealed that the ʃ-electron shell is closed . . . the ʃ energy per electron is 1.558."

SOURCE: Bochvar DA and Gal'pern EG, Hypothetical systems: carbododecahedron, s-icosahedrane, and carbo-s-icosahedron, Proc Acad Sci 209:239, 1973.

In the early 1970s in England, where the industrial revolution first started, David Walton developed methods for synthesizing long-chain polymers. This method was applied in acetylene chemistry. Long carbon chains and their *vibration-rotation dynamics* attracted Harold Kroto, who had been interested in research into molecular radioastronomy. In 1975, he conducted research on HC_5N and HC_2N molecules in space, and during the 1980s he became convinced that such ion-molecule reactions were taking place. It was at that time that the idea was born in his mind that carbon stars should be the key to understanding red giants.

1.5 Kroto/Smalley Research Team

In 1984, Kroto visited his colleague, Bob Curl, at Rice University in Houston, Texas. During his visit, he was introduced to Rick Smalley, who with his coworkers analyzed the spectrum of SiC_2. The result was unexpected: the molecule was not linear but triangular. Curl's suggestion that Kroto meet with Smalley and Smalley's SiC_2 spectral findings were key factors in Kroto's decision to visit Smalley at his laboratory. At the laboratory, Kroto saw the laser vaporization cluster-beam, a new apparatus which they had recently developed. He realized that with this device he could produce long carbon chains, and planned a joint project with Curl "in the hope that Smalley would also be interested."

A research group at Exxon had, in fact, already done something similar to what they planned to do. In 1984, they published an outstanding result: (1) both even and odd clusters for C_n, $1 \leq n \leq 30$; and (2) only *even* clusters C_{2n}, $20 \leq n \leq 90$. A new family of carbon clusters was discovered. In the interpretation of their remarkable experimental result, however, the Exxon research group was unlucky, concluding that it was ". . . evidence in favor of the linear chain structure for C_n as large as C_{24}." This exemplifies why theoretical and experimental research should go together more strongly than was the case in the Exxon group. Strong in the experimental realm but weak in theory and imagination, they lost primacy in this exciting field.

In August 1985, a year and a half after Kroto's first visit to Rice University, Curl telephoned him to say that the experiment they had planned to do with carbon was now ready to begin. Kroto packed his bags and travelled to the USA. The experiment started on Sunday, September 1, 1985--

NEW FAMILY OF CARBON CLUSTERS
Exxon Research Group

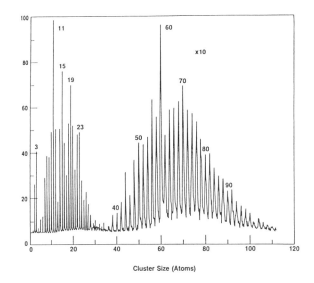

Cluster Size (Atoms)

This research group, Rohlfing EA, Cox DM and Kaldor A, did experiments in which they produced carbon clusters by laser vaporization of a solid graphite rod within the throat of a high pressure pulsed nozzle. The experiments showed the existence of two kinds of carbon cluster families: small clusters (odd) and big clusters (even) where peak of C_{60} was dominant.

SOURCE: Rohlfing EA, Cox DM and Kaldor A, Production and characterization of supersonic carbon cluster beams, J Chem Phys 81(7):3322, 1984.

THE FIRST FULLERENE PENTAGON
Kroto, Heath, O'Brien, Curl, Smalley

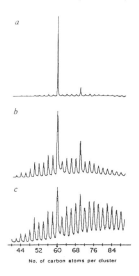

No. of carbon atoms per cluster

NATURE VOL. 318 14 NOVEMBER 1985

Their experiment showed the three different spectra, under different conditions. (a) was obtained by maximizing cluster thermalization and cluster-cluster reactions, (b) was obtained when roughly 760 torr helium was present over the graphite target at the time of laser vaporization, and (c) was obtained when effective helium density over the graphite target was less than 10 torr. They found that a remarkably stable cluster is one with 60 carbon atoms. They proposed that 60-carbon atom structure could be a truncated icosahedron, like a football (or a soccerball in the USA) and made a picture of a soccerball on Texas grass and published it in *Nature*.

SOURCE: Kroto HW, Heath JR, O'Brien SC, Curl RF and Smalley RE, C_{60}: Buckminsterfullerene, Nature 318:162, 1985.

three days after his arrival in Texas. Kroto was assisted in the laboratory by two research students, Jim Heath and Sean O'Brien. "Curl and Smalley paid frequent visits to the laboratory," too. "As the experiments progressed, it gradually became clear that something quite remarkable was taking place: As we varied the conditions from one run to the next, we noticed that a peak at 720 *amu* (atomic mass unit) behaved in a most peculiar fashion." This strongly indicated that there should be some kind of carbon molecule with sixty atoms, but what type of molecular structure it would have was unknown. This experiment was similar to that of the Exxon group, but was a better indicator of C$_{60}$, because the peak of 720 amu was 30 times stronger than other signals. After the experiment, on Friday, September 6, they had a group meeting. Curl mentioned that they should be "identifying the conditions under which the 720 amu peak was most prominent." He was right, because the value of the peak was the main difference from the Exxon result. The group in general concluded that the molecule should be some sort of spheroid. Kroto mentioned Fuller's geodesic dome and recalled that he, himself, had once built something similar for his children, but there were pentagons as well as hexagons in it. Of course, we now know that the key word was **pentagon**, but they didn't know that then. They borrowed a book, ***The Dymaxion World of Buckminster Fuller*** by R.W. Marks, from the Rice University library. Kroto had a number of different ideas, but one of them "came to mind several times, most vividly on this particular Monday. This was of a polyhedral cardboard star-dome which I had constructed many years before when my children were young . . . I itched to get my hands on it and even described its features to Curl at lunch time. I remembered cutting out not only hexagons but also pentagons." The next day, Kroto returned to the UK, and the following morning, Curl telephoned to say that Smalley had solved the problem because he had experimented with paper models based on the star-dome properties.

"It was midnight," said Smalley, "but instead of going to bed I went to the kitchen for a beer. Halfway through it I remembered Kroto's saying that pentagons might have been part of his children's geodesic dome. Maybe pentagons would be part of carbon clusters too; after all many carbon compounds are made of both five-member and six-member rings."

Sometimes history repeats itself. Watson and Crick had found a solution for DNA as a double helix using a *paper model*, and Smalley attacked his problem in a similar fashion. That night he went back to his desk, cut pentagons and hexagons from paper, and made the shape of a bowl. Step by step he made a nice hemisphere. His heart leapt because it "seemed beautifully symmetrical and quite sturdy." Every scientist who makes this extraordinarily beautiful structure must feel that his/her heart leaps stronger than before or feel some kind of "sweet" sensation. No one who sees this molecule as a 3-D structure could be indifferent toward C_{60}. "When mind discovers a generalized principle," said Fuller, "permeating whole fields of special-case experiences, the discovered relationship is awesomely and elatingly beautiful to the discoverer personally, not only because to the best of his knowledge it has been heretofore unknown, but also because of the intuitively sensed potential of its effect upon knowledge and the consequently improved advantages accruing to humanity's survival and growth struggle in Universe."

Curl's reaction was that beauty was fine, but what about the chemical reality of this kind of structure? What about double bonds between the atoms in such a structure? For an explanation of their structure, they asked for help. Smalley telephoned W.A. Veech, Chairman of the Mathematics Department at Rice University, and explained the kind of structure that he had made the previous night. Veech soon called him back and said, "I could explain this to you in a number of ways, but what you have got there, boys, is a soccer ball."

Which is why a picture of a soccer ball found a place in the November 14, 1985 issue of *Nature*. Kroto, Heath, O'Brien, Curl and Smalley, *first Fullerene pentagon (FFP)*, published a paper entitled, "*C_{60}: Buckminsterfullerene.*" This paper has become historically significant, although it contains some errors (like the position of carbon atoms from a symmetry point of view, and possibly some physical properties of the molecule). These misses are understandable for two reasons: (1) symmetry of the icosahedron was little known in science at that time (for example, Hammermesh, in his book *Group Theory and its Application to Physical Problems* [1972] said that "icosahedral group has no physical interest and no examples of molecules with this symmetry are known." Of course, as a classical physicist, he didn't know that viruses and some biological macromolecules possess exact icosahedral symmetry properties); and (2) they were in a hurry to publish and didn't take the time to adequately check their statements. The authors of this book are well aware of these kinds of errors and imprecision and welcome corrections and clarifications (to be incorporated into future editions).

The **first paradox** involving the C_{60} molecule occurred with the Kroto/Smalley research team. This *accidental synergy of mind* occurred as they spent a few exciting days trying to explain the structure of C_{60}, which had already been explained by Osawa and Bochvar/Galpern. The **second paradox** happened with Donald Huffman at the University of Arizona. He produced C_{60} molecules in his "graphite smoke" perhaps as early as 1973, but for certain from 1983. He was not aware of it, however, until 1985, and he didn't prove it until 1989.

1.6 Huffman/Krätschmer Research Team

K.L. Day and D.R. Huffman from the Department of Physics at the University of Arizona published a paper called "Measured Extinction Efficiency of Graphite Smoke in the Region 1200-6000 Å" in May of 1972. The graphite particles "were produced by striking an arc between graphite electrodes in helium." They also tried to produce graphite grains in argon but with no success. Both helium and argon are ideal gases, but for some reason the experiments only worked in helium. They used two sources of graphite: synthetic and natural. Some of the particles of graphite were irregular but compact, while another sort was "*quite spherical* with a mean diameter of about 500 Å." They proposed a spherical structure which is 50 times higher than the actual C_{60} molecule known today, while calculations by other scientists suggested that the radius of spheres should be 150 Å.

Huffman made some trivial changes in the experimental procedure together with Wolfgang Krätschmer (Max Plank Institute of Heidelberg) in 1983, the same year that Fuller died. Norbert Song assisted them in some part of the work. "There was no outstanding difference in the sample production technique from that used earlier, but we investigated many variations of the parameters: gas composition, pressure, types of arc, and position and configuration of the collecting surfaces." That year they did UV spectra and Raman spectra for some different forms of carbon smoke. Based on these data they defined "**the camel sample**" effect in spectra, but they concluded that it should be some sort of "*junk*." Unfortunately, the word "junk" was frequently used in their interpretation of their experimental results of "graphite smoke." At the time, "*graphite smoke*" wasn't a high priority project in either Tucson or Heidelberg.

The paper of the **FFP** in 1985 was a bomb hidden in the Huffman/Krätschmer unknown research of "graphite smoke." From that time they began to consider that their "junk" could be relevant to the C_{60} molecule.

F:1-7

FULLERITE
A new solid C$_{60}$ form of carbon

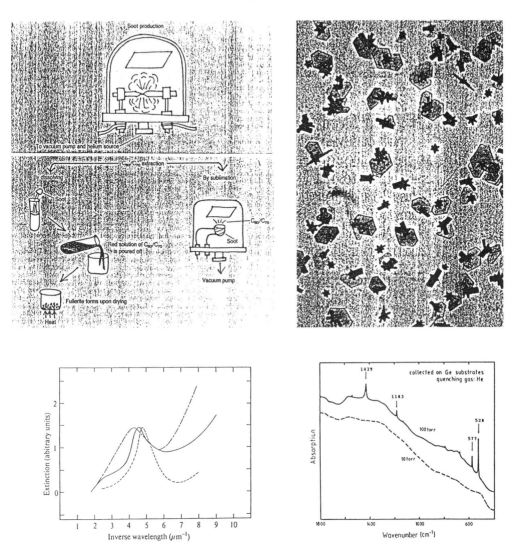

(1) Huffman method of producing solid C$_{60}$ and C$_{70}$ which he has started to develop until 1970s. For many years he and his coworkers believed that they produced "junk" material. Today it is "carbon gold." (2) The first crystals of solid C$_{60}$/C$_{70}$ (plates, rods and stars). A new science, Fullerite was born. (3) Measured extinction for "graphite smoke," and graphite spheres calculated in 1973. (4) Infrared absorption spectra of soot produced at different quenching gas (helium) pressures. These four peaks strongly suggested that C$_{60}$ molecules are produced in Huffman/Krätschmer "graphite smoke," because theory predicted that C$_{60}$ has only four infrared modes.

SOURCE: (1)   Day KL and Huffman DR, Measured extinction efficiency of graphite smoke in the region 1200-600Å, Nature Physical Science 243:50, 1973.
(2)   Huffman D, Solid C$_{60}$, Physics Today 11:22, 1991.
(3)   Krätschmer W, Lamb LD, Fostiropoulos K and Huffman D, Solid C$_{60}$: a new form of carbon, Nature 347:354, 1990.

Huffman heard the news of the FFP paper from his graduate student (now Assistant Director of Arizona Fullerene Consortium), Lowell Lamb, which he had read in the *New York Times*.

Huffman spent the next two years producing macroscopic quantities of C_{60} from "graphite smoke" to submit a patent. But "when time came for serious discussions with the patent attorney," said Huffman, "I was unable to reproduce the **'camel sample.'**" However, the following year (1988) in both Tucson and Heidelberg, independent production of the "graphite smoke" under varying conditions began (with assistance from Lowell Lamb and Bernard Wagner, respectively). Soon "camel humps" were again recognizable in the ultraviolet domain. The researchers compared four infrared-active modes which were predicted for the C_{60} molecule (based on symmetry properties of C_{60} molecule) with their result of the "camel humps." The four infrared bands in the ultraviolet domain of "camel humps" and C_{60} molecule were the same. **Eureka!** Huffman had originally started to investigate "graphite smoke" to solve a problem in astrophysics, and though he found a "forest" of carbon clusters, he wasn't looking for C_{60} molecules. Nevertheless, he did find C_{60} and saw a *new world* of solid carbon in front of him. This pattern of innovation is really unusual, as many other great discoveries have been equally serendipitous (e.g. Kekulé, Gauss, Lomonosov, Tesla, and many others). Little did Göethe realize when he wrote his poem, "Gefunden," that it would someday be relevant to researchers (Huffman, Kroto, Koruga) in the field of Fullerenes from different viewpoints:

> *Ich ging im Walde*
> *So für mich hin*
> *Und nichts zu suchen,*
> *Das war mein Sinn.*
>
> *Im Schatten sah ich*
> *Ein Blümchen stehn*
> *Wie Sterne leuchtend*
> *Wie Äuglein schön.*

(In free translation, the poem says: "I was walking in the forest, not looking for anything. In the shadow . . .")

The second reason for Huffman's **eureka** was his personal feeling; a "dream" of his mother had become reality. She had known in some way that her son Donald would discover something important based on the number 60. As an elderly lady, she was blind, but she counted her steps and had told Donald that the number 60 was her "magic number." In memory of his mother, Huffman gave the first lecture of his course, **"Physics of Fullerenes,"** at the University of Arizona in a large auditorium filled with professors and students on August 26, 1992.

Their results (the Huffman/Krätschmer Research Team) were presented in 1989, but some scientists were skeptical of their findings. They spent a year solving the problem of how to separate the Fullerenes from the soot and make it into crystal form. A solution was found in Heidelberg: wash the soot in *benzene*. Huffman had written in his 1973 paper, "carbon particles (soot) was generated by burning **benzene** in air and collecting the material on a substrate." They were able to produce about *100 mg* of extracted solid C_{60} as a new form of carbon per day. They were the *first* to see a new *crystal* form of C_{60} molecules, and a *new science* of Fullerene was born. In September 1990, they published a paper in *Nature* about the basic properties of solid C_{60}, or **Fullerite**, as they called it. The story of how they discovered what kind of crystallographic symmetry group solid C_{60} possesses is a very interesting detective plot; we hope that Huffman and Krätchmer will publish it soon.

Finally, in 1991, Huffman/Krätchmer submitted a patent for the production of Fullerenes through the Office of Technology Transfer at the University of Arizona in Tucson. The Materials and Electrochemical Research (MER) Corporation in Tucson has the world exclusive license of the Huffman-Krätschmer patent applications for composition of matter for Fullerenes and the synthesis extraction process.

1.7 Loutfy/Withers Research Team

MER was formed in 1985 by Dr. J.C. Withers and Dr. R.O. Loutfy. Dr. Loutfy is the president of the company, while Dr. Withers is Chief Executive Officer. Dr. Loutfy has 17 U.S. patents, and Dr. Withers has 25.

F:1-8 **MER CORPORATION**

Dr. Withers (in middle) and Dr. Loutfy (right) shake hands with President of Taiwan, Don Hue Lee, during their business trip for Fullerene production in January 1992.

The first prototype of Ni/C_{60} rechargeable batteries, whose performance is superior to that of Ni/Cd batteries.

They are both principal investigators of projects in the field of Fullerene technology sponsored by NASA, the U.S. Army, and other companies. The company employs a team of about 35 researchers for technological development and production of Fullerenes. Their primary goals are the production of Fullerenes, technology development, and applications of Fullerenes.

Technology and applications provide a link between MER and the Koruga/Hameroff Research Team at the University of Belgrade and the University of Arizona.

1.8 Koruga/Hameroff Research Team

Mind synergy and the implementation of knowledge from **nanobiology** to **nanotechnology** are basic characteristics of this research team. Koruga and Hameroff started independently during the 1970s to do nanobiology research in the field of microtubules. Koruga proposed to do research on Fullerenes in science and nanotechnology, while Hameroff (stimulated by Conrad Schneiker) attempted to develop scanning tunneling microscopy (STM) as a basic tool for nanotechnology. Koruga's link with today's Fullerene carbon molecule (C_{60}) dates from the early 1970s. Which brings us to *the third accidental event-- serendipity* in the field of Fullerenes. Koruga's experience might be more understandable in the context of the book **Serendipity**: *Accidental discoveries in science* by R.M. Roberts.

1.9 Koruga's Serendipity Experience: Carbon Dream

Beginning in May 1973, all of my conscious scientific activities have been focused on understanding the fundamental relation of the Man-machine system. I studied in earnest both biological and technical sciences, with the evolution of the creation of Man and machine and some kind of transformation of Man into machine from the energetic-informational aspect being my main preoccupation. My research led me in the direction of bioelectronics; however, the question on which fundamental principles of future information machines will be based has so far remained unanswered.

For years, my whole being has been occupied with the search for the answer to this question. Whether all this conscious activity influenced the appearance of a dream is unknown. One night in October 1978--I vividly remember, as if it only happened yesterday--I dreamed that I was again at my birthplace, in a part of the village where, during my childhood, friends and I often played soccer and made bubbles from soap found in the nearby washtubs. The place was called Bare (English translation: Puddles) and was one of only three locations in the village where water could be found. Large washtubs were set up there to hold the water, and local women came there to do their wash. We children would use the occasionally-forgotten washing soap to make bubbles, which would then float all around the little clearing. This clearing was surrounded on all sides by heights and hillpeaks.

In my dream, there suddenly appeared **two red suns** in the clear and cloudless sky above the surrounding hills. One of the suns was in **the east** while the other was in **the west**. The sun in the eastern sky was motionless at a point one-third of the field-of-view from the east, toward the center of the viewfield. The sun in the west was located about one-fifth of the field-of-view from the west toward the center and was slowly moving toward the sun in the eastern part of the sky. All of a sudden, the two suns touched. Although I was eagerly anticipating what would happen then, at first nothing happened. Then, to my great surprise, the sun which had been moving from the west continued to move. At the moment when it **covered** the other sun, they both turned to blinding bolts of lightning. Lightning struck the summit of the tallest peak dwarfing the village, and instead of a thunderclap, a godly voice resounded, "Carbon, 13, 18, 24, 2 becomes 3!"

I awoke and at once wrote down the contents of the dream I so vividly remembered, feeling it contained some very important message which at the time I had not understood. This dream brought me a sense of well being, and at the same time a feeling of *beauty* and *sublimity* that was to stay with me for the next several days as the events of my life slowly buried this experience somewhere in the deep recesses of my mind.

About three weeks after I dreamed of the two suns, my sister (Leca), who lived in a nearby town, about 100 km from my home, paid us a visit. After we finished our meal, we sat at the dinner table and talked. Without any prompting from me, she began the following story: "Brother dear, last week I went to visit Mother, and we talked about all sorts of things. For some time now, I have wanted to ask her the name of that hill at *Padalište* (the place where something fell from the starry heavens to the earth; before the people

F:1-9   SVETLANA'S MIND SYNERGY
                                      Poetry - Art - Science

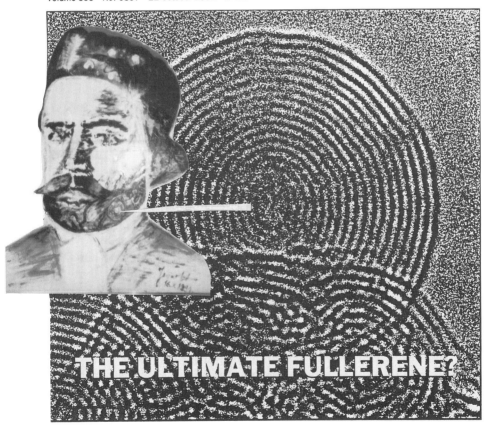

Svetlana Janković illustrated the five-fold symmetry of C$_{60}$, Einstein's relation of mass and energy through light, and Cantor's middle-third set as a "snail" on this portrait of Petar Petrović Njegoš [1813-1851] (October 16, 1991). The same form of a "snail" or "onion" was found as the form of Fullerenes. The smallest circle on the picture of the "snail" represents C$_{60}$ (*Nature*, October 22, 1992). Note: The first transmission electron micrograph of giant Fullerene as concentric shells was done by S. Iijima (J. Cryst. Growth, 5, 675-683, 1980), but she (Svetlana) didn't know that, and Iijima didn't recognize C$_{60}$ as a "spherical particle of carbon graphite" at that time.

believed when they see comets on sky that stars fall down on earth), where as a girl I herded sheep. I used to know the name of the hill, but that was back when I was a child and I have since forgotten. I asked Mother, and she told me its name. Did you know that it's called *Ključ* (English translation: **Key**)?"

A shiver ran through my entire body, and my dream of three weeks before sprang to my consciousness. The peak struck by lightning in my dream and the one my sister was asking about were one and the same. A bolt of lightning now burst inside my mind. The **key** to the **solution** of the fundamental principle of the Man-machine system, or biological life as we know it and future artificial life, was **carbon**. "Bioelectronics," was my first thought, but that was in contradiction with the development of electronics based on silicon. Because electronics based on silicon was rising toward its zenith, there was no room for any scientific discussion of carbon as a material for electronics.

The dream's main message was becoming clear: life is based on the carbon atom as symbolized by the sun in the east--a solution already achieved in my dream--while future machine artificial life will also have to be based on carbon, as the sun in the west. But what was the meaning of the numbers I heard in my dream? Their significance was not clear to me at that time, and I must admit that not all of them are comprehensible to me now. My sister's visit brought back memories of the beauty and sublimity I had experienced in the dream.

During the next several days, I began writing about what I experienced in the dream. I heard that the next year (1979) the 9th World Congress of Aesthetics on the theme of creativity was scheduled to take place in Dubrovnik. I called Professor Milan Damjanović, one of the organizers, and told him that I would have something interesting to contribute to the theme of creativity in the context of Kant. I briefly outlined what I would write about to Professor Damjanović without revealing the contents of my dream. I called the paper "Creativity as an experience of the beautiful and sublime." He agreed that I should prepare the paper, which I accomplished in a short time. While I was writing the paper based on my dream, three things became crystal-clear: carbon, ancient Egypt and axis of 5-fold symmetry will determine the fundamental principles and future of machines--artificial life. In my paper, I wrote: ". . . Man and Nature through Machine, the New World will be founded with the New Man incorporated. The New World will have the symmetrical base of axis five." The presentation of the paper at the Congress was very exciting for me. The paper was received as very interesting, but any encouraging feedback was minimal. My state of "excitement" lasted for the next several months, after which the experience began slowly to recede into the depths of memory.

From time to time, the dream would return to my consciousness, as one of the most powerful experiences of my life. At the end of 1978, I began serious research into the field of symmetry, resulting in 1984 in the publication of my book *Qi Engineering*, in which I described the existence of three systems of symmetry: crystallographic, Curie and 5-fold symmetry. I ended the book with the following words: "Qi engineering as an integral part of Nature, Man and Machine represents a scientific discipline which reassesses the entire existing view of the world of reality on the basis of the laws of symmetry and develops a methodology and direction of its action on the basis of the dictates of these laws." Research into the field of symmetry has proven to be of great significance, because it has shed new light on the explanation of some biological-informational mechanisms. This led to new approaches in Neurocomputing, which I reported at the NATO conference in Tucson (1991) and later published in the journal, *Nanobiology*.

In 1984, I met Stuart Hameroff from the University of Arizona. He was working on the simulation of information processes in microtubules, while I was researching structure, symmetry and the code system of microtubules. To experimentally test our ideas on the structure and dynamics of microtubules, he suggested that we begin research in the field of STM (Scanning Tunneling Microscopy). Anticipating the future significance of STM, on the suggestion of Conrad Schneiker, Stuart had begun research in the field of STM in 1983 at the University of Arizona. Shortly thereafter, the first STM was assembled at the University of Arizona.

Grasping the importance of STM to nanotechnology, upon my return to the University of Belgrade in 1985 I founded the Molecular Machines Research Center (MMRC). That same year, I read an interesting paper in *Nature* by the Kroto/Smalley research team about C_{60}. I read the paper, thought it interesting, and placed a photocopy in my files, as the paper did not provoke my immediate curiosity. During the next several years, I equipped the MMRC with STM/AFM (atomic force microscopy), a neurocomputer, neurochips, transputers and other equipment. I formed a research team that mainly studied bioelectronics, nanotechnology, microtubules, clathrin and followed the results of research in the field of C$_{60}$. We also worked on research into the implementation of biology to technology (biological and artificial neural networks). With the assistance and encouragement of Professor Ljubomir Grujić, I introduced a new course in Bioautomatics at the University of Belgrade during 1987. One of the subjects of study in this course is bioelectronics and information principles based on the *Golden Mean* in biological (microtubules, clathrin, etc.) and technical systems (artificial microtubules in the future). However, it was the work of Harold Kroto published in *Science* in November 1988 entitled "Space, stars, C$_{60}$ and soot" that attracted my attention. After that, I began studying the increasingly interesting C$_{60}$ molecule from the aspect of symmetry but still did not tie it in with my dream. It now seems to me that during the period of 1986-1991 I may have "completely" forgotten about the dream.

During the spring of 1991, by pure chance, Professor Miloje Rakočević of the University of Niš, came to the MMRC carrying a bundle of papers with his calculations and proofs that master authors like Göethe, Tolstoy . . . Njegoš wrote masterpieces based on the principles of the Golden Mean. At that time I was teaching Golden Mean properties of microtubules in my course, Bioautomatic Control, and intensively working on the symmetry of C$_{60}$. Knowing that the Golden Mean is the basis of 5-fold symmetry and the C$_{60}$ molecule, and as a connoisseur of poetry and prose, I was interested in his ideas. At first skeptical about the possibility of applying the Golden Mean to literature, I began to realize as the interview progressed that Professor Rakočević may have uncovered something, but I had to have time to verify.

We agreed to meet again in two or three days time. During those couple of days, I reread "Ray of the Microcosmos" be Petar Petrović Njegoš (one of the greatest Yugoslav poets, prince of Montenegro, philosopher and bishop; born 1813, died 1851), one of the examples Professor Rakočević had spoken of as having been written in accord with the Golden Mean. My surprise was complete. The poem, "Rays of the Microcosmos," about Man and his relation to his God (Creator), his destiny, was really written according to the Golden Mean. I experienced both shock and joy. Was it then possible that masterpieces by Göethe, Shakespeare, Tolstoy, Njegoš and others were written in accord with the Golden Mean, and were they masterpieces because they were written according to the Golden Mean? When again I met with Professor Rakočević, we decided to organize a scientific symposium about the works of Njegoš in order to present these ideas to a wider audience. I asked one of my research associates, Svetlana Janković, M.D., who also happens to be a talented amateur painter and fan of the poetry of Njegoš, to paint a portrait of Petar Petrović Njegoš.

At the MMRC, Svetlana was researching C$_{60}$ from the aspect of applications of Fullerenes in medicine. The first time I spoke with her about C$_{60}$, she experienced a sort of "sweet headache" and felt something strange which she could not explain. She brought me the finished portrait several days later. I was surprised at her **synergy** of art and science. She explained the portrait to me part by part as she had envisioned it, each part conveying a particular message. The portrait was actually her vision of Njegoš' poem, "The Ray of the Microcosmos," which was based on the principles of the Golden Mean. The detail that caught my attention the most was when she explained that she painted the *eyes* as *two red suns*. I don't know why, but I asked her if maybe they were the sun and moon. She answered in the affirmative, confirming that they were two suns, leaving me with a strong impression of awe. Studying the portrait further, I noticed another detail that fascinated me, but a detail that Svetlana had completely omitted in her explanation. I asked her what she saw in the region which she called a "snail," where she had painted Einstein's and Cantor's equations, which she had already mentioned. She started again to explain what she had tried to convey with this or that detail, but again omitted the particular detail I had noticed. This only strengthened my opinion that this detail had not been painted consciously, but *was created* during the process of painting. Finally, I asked her if she saw the number *five* as the result of the *synergism* of the "snail" (C$_{60}$), Einstein's and Cantor's equations. She pointed to the number five which she had consciously painted but again did not see the number *five* which represented the nature of the portrait. Then I pointed out to her the detail on the portrait which represented the number *five*. She was uncertain whether this really was what I had told her, but finally, almost to herself and under her breath, she mouthed the word "interesting." I had the impression that she did not at that time give much attention to this as a detail (link between Njegoš, Fullerene, Einstein, and Cantor) of importance, mainly because this detail had been created in the process of painting outside of her consciousness. It was that detail, however, that had been created as a result of **mind synergy** on the *unconscious level*, something that is characteristic of creative minds giving *aesthetic value* to any work, to masterpieces.

After the scientific symposium on the works of Njegoš, we continued

F:1-10

Dreams we usually feel as nightmares. We are like in prison.

If we try to understand the dream, we should play the friendly game with him.

Koruga's Carbon Dream as Serendipity

At once you will discover a great potential of unconsciousness, after you analyze a number of your dreams.

You will recognize that something what was for your nightmare can be salvation.

If you combine in right way your consciousness and unconsciousness, you will find right solution, and escape from "prison."

Your mind will be transformed, as synergetic one, in true sense. Life wings (UFO) is born.

with our daily research into microtubules, STM, bioelectronics and neurocomputing. We prepared an abstract on molecular electronics for the First World Congress for Electricity and Magnetism in Biology and Medicine. My idea was to create artificial microtubules based on C_{60} as a result of our research into nanobiology--biological microtubules and clathrin. We finished the abstract and sent it off for the Proceedings. We continued the paper, and as the final touches were being made, I received the November 7, 1991 issue of *Nature*. Leafing through the issue, I came across the article "Helical microtubules of graphitic carbon," which at once grabbed my attention.

I went out of my office into the room where my associates work. I showed them the article and was distressed that we had not already published our paper on artificial microtubules based on C_{60}. When I read the whole article, however, I saw that our idea was a bit different, which calmed me down considerably. After reading the article, I went back to its beginning to see who the author was, since the title and contents had been of more interest to me initially and I had overlooked the authors, only remembering that they were from Tsukuba, Japan. I remembered this easily enough as I had given a lecture on microtubules there in 1986 when I visited the Electrochemical Laboratory in Tsukuba. I then remembered that the lecture was attended by a researcher from NEC Corporation. I looked through my album of the 1986 visit to Japan and saw that it was not the same researcher. All of a sudden, I began to repeat to myself "Japan, Japan . . .," then the Japanese flag and its rising sun came to my eyes. Mechanically, I looked up at the portrait of Njegoš which now hangs in my office, and I looked at those eyes painted in the form of two red suns and at that moment I felt as if I had just awakened from some kind of dream state. The picture of the red sun from the Japanese flag and Njegoš' two eyes as two red suns came together in a single picture and brought back memories of the dream I had had in 1978.

Right at that moment, *something changed* in my very being, something that is very hard for me to describe; at that moment, I felt things had fallen into their places. What is even more interesting is that exactly 13 years had passed since I dreamed my "carbon dream." My first thought was that the sun in the east in my dream symbolized Japan; however, when I made a detailed analysis of my dream and the article written by Iijima, I saw that something was missing. I took out my C_{60} file and began reading the articles I had collected. The earliest article I had was written by the Kroto/Smalley research team in 1985. In the righthand corner, I noticed a soccer ball. That associated me with the place from where I had watched the two suns in my dream. This could only mean that there had to be some connection between the soccer ball and Japan. However, I could make no connection, although I knew a solution had to exist. I went back to writing the article for the Congress and then when I least expected it, the answer presented itself. In the paper I had already written,   ". . . Having in mind C$_{60}$ as the third known form of carbon (graphite, diamond and buckyball) has the same symmetry properties as **clathrin** we propose to synthesize C$_{60}$ as helical artificial microtubules with information properties similar to natural MT." Although I had already written that, I only then became aware that the two identical suns in the sky of my dream represented the **same symmetrical structures**. The power of symmetry in science is symbolically the same as the power of the sun in nature. C$_{60}$ and clathrin have the same symmetry--icosahedron. If I correctly understood my dream, it had to mean that a Japanese researcher discovered the symmetry (structure) of clathrin. I quickly reached for my clathrin file and began to leaf through the papers there. There were about 20 papers on clathrin there, but I found nothing there that could confirm my hypothesis.

During the next several days, I returned to my regular duties and began preparing for my trip to Tucson. On the trip to the U.S., I brought a list of papers published on the subject of clathrin. Several weeks after I had arrived in Tucson, I took my list and began looking over the papers. I decided to start with the "source" paper on clathrin. The first time clathrin was mentioned in the title of an article was a paper by Barbara Pearce published in 1976. I went to the library and found the paper. I recognized it as a paper I already had in my files at the MMRC in Belgrade. I reread the paper and only on the very last page in the list of literature did I find something of interest--two Japanese authors, Kanaseki and Kadota. Right away, I looked up the reference (J Cell Biol 42:202-220, 1969) and with the anticipated "surprise," I found that in the article about clathrin the two Japanese researchers had included a picture of a **soccer ball**, just like Kroto/Smalley, to explain the structure of clathrin. For some time, I sat in the library thinking about the nature of human beings, symbolism, and the power of the human unconscious.

Even before this event, my intuition told me that things happen according to some logic we mortals could not yet fathom. This event influenced me to pass from a state of intuition into a state of knowledge of the **sychronicity** of events in **space** and **time** and of the entire human race as a unified system which acts according to principles very similar to those we find

F:1-11

BIO-FULLERENE
Clathrin

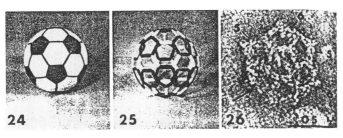

In 1969, Kanaseki and Kadota, two Japanese scientists from the University of Osaka, discovered that clathrin (very important protein in each cell) has icosahedron symmetry and looks like a soccerball.

New Intelligence

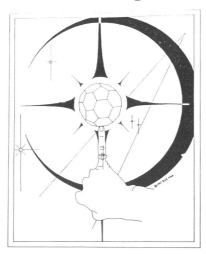

Buckyballs & BioElectronic Life

Dr. **Djuro Koruga** of the University of Arizona **Advanced Biotechnology Lab** is pursuing an age-old dream: the synthesis of bioelectronic extensions to human life. Djuro's research focuses on creation of buckyball (C60) cytoskeletal arrays for **subneural neurocomputing**, **intelligent control** of cytoplasmic organelles and **bioelectronic interfacing**. This could someday bring whole new worlds of thought to our fingertips or *into* our fingertips.... Goodness! Gracious! **Great balls of fire!** Bucky would have been so proud.

IEEE Neural Nets Group1 Meeting
Thursday, April 16, 1992
6:45 PM COB 152
Arizona State University

The first lecture in the field of bioelectronics after 13 years from Koruga's "carbon dream." Rick Alan illustrated this (two suns are overlapping with C_{60} in middle) and wrote text without knowledge of "carbon dream."

in science as deterministic chaos.

Not long after this "carbon dream"-related event which occurred in Tucson, I met Rick Alan of TRW Corporation. Rick had invited me to speak to the IEEE Neural Nets group at Arizona State University in Phoenix. The invitation was based on what Stuart Hameroff had told him about my research in the field of Fullerenes and biomolecular information processes, and he chose the title of the lecture, **Buckyballs and Bioelectronic Life**. Neither Stuart or Rick knew anything at the time of my dream about carbon as starting point for my research in bioelectronics; they only read about it in the draft of this story for this book.

At the lecture, I explained my vision of the basic concept of interpreting multi-dimensional spaces ($N \leq 5$) on the basis of symmetry and "artificial life" based on C_{60} and 5-fold symmetry. Several days after the lecture and a few months more than 13 years after my carbon dream, I held C_{60} in my hand as a product of western civilization, symbolized by the sun in the west which was slowly moving. Part of the sample came from Huffman's Laboratory and the rest from MER Corporation in Tucson. I decided to send the samples to my MMRC in Belgrade for STM analysis. I sent the samples off by DHL so that three days later they were in my laboratory. Several days later, Jovana Simić-Krstić and Mirko Trifunović, two of my research associates, telephoned to say that they had obtained STM scan images of C_{60} at **atomic resolution**. Minutes later, I held in my hand a fax with three images that the duo had obtained. I showed these images to a few researchers at the University of Arizona.

As the images on the fax transmission were not very clear, we decided to wait for the photographic images to arrive before making any comments. Several days later, the photographs arrived by DHL. They were fantastic--on one, you could see clearly energy concentrated in pentagons; on another, clear images of 5 and 6 member rings of carbon atoms (or orbitals); while the third showed the *Ih* symmetrical basis of C_{60}. I called Huffman's laboratory to tell them that we had obtained original images of C_{60} at atomic resolution and that I wanted their comments on them. As they were overbooked that day, I did not get a chance to see them. So I called MER Corporation and set up an appointment with them right away. Our discussion was interesting; it was the **first time** the human eye had looked at the molecular structure of a new molecule with **atomic resolution**. MER Corp., as the producers of Fullerenes using the Krätschner/Huffman patent, attracted my interest because they are a well-equipped enterprise led by two experienced men in the field of

F:1-12

Proposed model of Fullerene C$_{60}$ as "Molecule of the Year" (*Science*, **20 December 1991**). Five months later, on April 24, 1992, scientists made scanning tunneling (STM) images of C$_{60}$ very similar to the proposed model. [The STM image resulted from a collaboration between the Koruga/Hameroff Research Team (University of Belgrade, University of Arizona) and MER Corporation. The image was done in the Molecular Machine Research Center at the University of Belgrade by Jovana Simić-Krstić and Mirko Trifunović.]

F:1-13

The first international promotion of nanotechnology based on the C$_{60}$ molecule at the 4th International Symposium on Bioelectronics and Molecular Electronic Devices, December 1992, Miyazaki, Japan. D. Koruga is in the first row, eighth from the right. Next are: Dr. M. Aizawa, Meada (FED), MITI representative and G. Matsumoto.

electrochemistry.   Our discussion soon turned to possible scientific, technological and production cooperation.   We reached a preliminary agreement, and I spent the next several months writing a detailed proposal for cooperative research.

In August 1992, I spent several days in Belgrade. I took with me more research material of C_{60} to be analyzed at the MMRC. My graduate student Mirko Trifunović and I worked intensely for several days on the STM. Svetlana Janković also assisted, but only intermittently, because she also had to work in the medical clinic. One of the results we obtained included C_{116} as the C_{60} "strong" dimer. We also recorded scans of several other forms of Fullerenes: with negative current, needles, nanotubes, ... On my return to Tucson, I continued cooperation with MER Corp., began a new year of lectures at the University and began work on a book titled *Fullerene C_{60}: History, Physics, Nanobiology, Nanotechnology*, which you now have before you. Following the discovery of a new form of **carbon molecule** by the Kroto/Smalley research team, this book proposes to give both direction to **nanotechnology** and a possible answer to the question, "What purpose(s) will Fullerene C_{60} have?" The opportunity to give a possible answer came soon in November of 1992, from Japan. Dr. Masuo Aizawa, as organizing chairman of the 4th International Symposium on Bioelectronics and Molecular Electronic Devices, sponsored by the Research and Development Association for Future Electron Devices (FED) and supported by Japan's Ministery of International Trade and Industry (MITI), invited me--together with Shuku Maeda (FED)--to give a plenary lecture at the conference on Fullerene C_{60} and its applications in molecular electronics. Most of the material which I presented at the conference is included in this book.

1.10 Geological Fullerenes

These three accidental events (Kroto, Huffman, Koruga) are not the only surprises in the field of Fullerenes. At a time when scientists thought that the synthesis of C_{60} and other Fullerenes was possible only in laboratories, or maybe in stars, an article in the June 10, 1992 *Science* was like **a bolt out of the blue**. Buseck and Tsipursky (Arizona State University) and Hettich (Oak Ridge National Laboratory) reported that "By means of high-resolution transmission electron microscopy, both C_{60} and C_{70} Fullerenes have been found

F:1-14   **FULLERENES FROM THE GEOLOGICAL ENVIRONMENT**
A bolt out of the blue

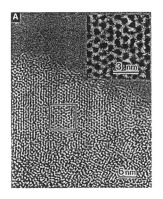   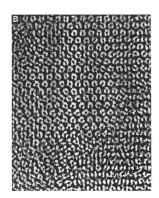

(1)   High-resolution transmission electron microscopy (HRTEM) image of the
        Fullerenes in a carbon-rich Precambrian rock from Russia.

(2)   Image of synthetic Fullerenes with HRTEM at the same magnification as in (1).

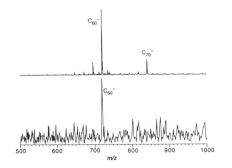   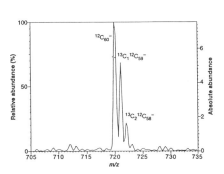

(3)   The laser-desorption spectrum shows the presence of C_{60} and C_{70} Fullerenes.

(4)   Also, the thermal desorption measurements verify the presence of C_{60} molecules.

SOURCE: Buseck PR, Tsipursky SJ, Hetich R, Fullerenes from the geological environment, Science 257:215, 1992.

in a carbon-rich Precambrian rock from Russia." That means that Fullerenes should be very stable, because Precambrian rocks are about a billion years old. Of course, we should be careful about setting the age of geological Fullerenes, because the Fullerenes in Precambrian rocks could be the result of secondary products of the geological environment or some cosmic accident. But in any case, if control experiments confirm previous report, "**geological Fullerenes**" are very old. This preliminary report raised a new question about the synthesis of C$_{60}$ and other Fullerenes.

1.11 Fullerenes: From Two to Three

For the end of the introductory story about Fullerenes, let's go back to Bio-Fullerene as a *fifth* accidental event in this field. For classical physicists, the "world" of physics and the "world" of biology are different. But when you ask them what the difference is between the two worlds, they are unable to give you an exact answer. If they could, they would be very close to knowing "what **Life** is." The relationship of Fullerene classical physicists to Bio-Fullerenes today is similar to the statement in science about the Sun and the Earth movement before the Copernicus era. Before Copernicus, the Earth was considered the *center of the universe*, but afterwards the Sun became the center of the planetary system. According to our initiation, Bio-Fullerene looks like **the King** of the *Fullerene world* (as five-fold symmetry one), while *the mother Queen* is somewhere in the sky, and today Fullerenes in our laboratories--and tomorrow in a new kind of machines--are like kids. And we know that kids one day become adults. This book is the first lesson about initial "kids'" steps; from Fullerene science to Fullerene nanotechnology.

F:1-15

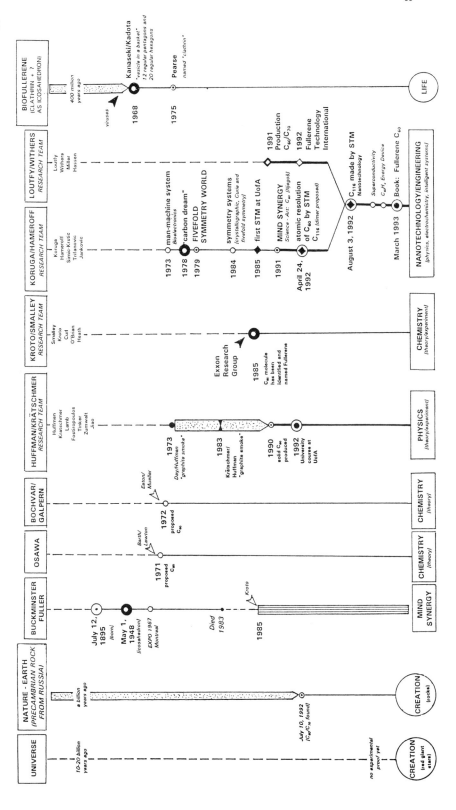

CHAPTER 2

THE ICOSAHEDRON
Symmetry and Golden Mean Properties

2.1 Short History

In the Platonic era of ancient Greece, four elements were postulated: *Air*, *Earth*, *Water* and *Fire*. These elements were represented as **octahedron**, **cube**, **icosahedron**, and **tetrahedron**, respectively. The Universe was imagined as a **dodecahedron**, related to the *icosahedron*. This is apparently the first time that a systematic categorization of geometrical bodies representing *symmetrical classes* was utilized. The Greeks used pure *visualization* in their thinking about their four elements and the Universe. We have a similar concept today: the four forces (*gravity*, *electromagnetic*, *weak*, and *strong*) and *superstring* as a theory of everything (**TOE**). This concept is also based on symmetry, but in a more complicated form than was that of the Ancient Greeks.

The major properties of the **icosahedron** are that it has 12 vertices, 20 faces and 30 edges, with the vertices possessing the property of a *5-fold symmetry axis*. The **pentagon** was also often used as a concept of the Universe in such other ancient civilizations as China, India, and Egypt.

The ancient Chinese believed that behind the visible order of Nature lay *Dao*, the universal cosmic principle which brought about the existence of *Yin* and *Yang*. The interaction of these two opposites created the **five** elements: *earth*, *fire*, *wood*, *water*, and *metal*. From this pentagon, all things arose. Linguists such as August Schleider included the Chinese language as a root language, because relationships among words are determined by a certain word order and, therefore, the Chinese language, which is pictoral-phonetic, is complementary to the world of symmetry (crystals). Analysis of Chinese characters (*Hanzi*) can lead to an understanding about mankind at a time when human thought did not position itself in relation to something else as a *subject* does in relation to an *object*, but instead considered the existing reality directly. Owing to the preservation of the unity of form and content in Chinese characters, their analysis leads to an unveiling of the fundamental meaning of the **pentagon**, as the *synergy* of **five** elements. The crown of Chinese mind synergy based on the pentagon is the ancient character which expressed *personality*, i.e. the personal pronoun, "I," *Wo* in Chinese, which was written as meaning "*I am the speech of five*." The basis for a five-fold symmetry world, as we will consider with Bio-Fullerenes, was first found in the ancient Chinese language. Because of the preservation of picto-phonetic unity--that is, of the unity of the form and the content--Chinese characters and Chinese

F:2-1   PLATONIC SOLIDS

TETRAHEDRON
Fire

OCTAHEDRON   **ICOSAHEDRON**   **CUBE**
Air   *Water*   *Earth*

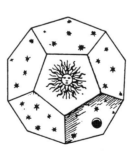

DODECAHEDRON
the Universe

The system classification of main four elements in ancient Greek's world. Water has double role: (1)
Earth-Air circulation (water in rivers, oceans . . . on Earth, clouds in air, and raining–principle of
synthesis of hydrogen and oxygen), (2) Fire-Universe relationship (when water separates into hydrogen
and oxygen, then hydrogen gives fire and oxygen support hydrogen's burning–principle of analysis).

F:2-2   ANCIENT CHINA AND JAPAN

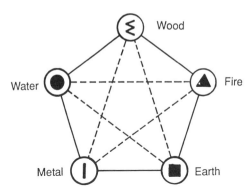

Ancient Chinese concept of five elements and their representation as pentagon (full lines) and star (dashed lines). Universe in pulsation (two extreme states) has one of these states. Scientists have found a similar result for the C_{60} molecule for its HOMO (highest occupied molecular orbitals) and LUMO (lowest unoccupied molecular orbitals).

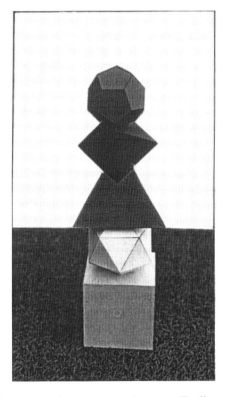

A Gorinto, a five-storied small pagoda, consists of five blocks. On each block a Sanskrit letter or Kanji character naming each of Indian five elements: earth, water, fire, air and heaven. Earth and water give one category, while fire, air and heaven give another category. Ratio(s) between these two categories could be 3/2 or 2/3, but the product is 1.

civilization represent a potential treasurehouse of knowledge about Nature and man. A similar situation occurs in Japan, because of the three Japanese alphabets (Kanji, Katagana, Hiragana), one--Kanji--also possesses the picto-phonetic properties of unity. A small, five-storied pagoda known as a *Gorinto* is a symbolic tower for Buddhists in Japan. The pagoda consists of five Platonic solids, each representing one of five Indian elements: *earth*, *water*, *fire*, *air*, and *heaven*.

In the *Vedas*, the oldest Indian written documents, which date back to two thousand years B.C., gods are generally rendered as personifications of the forces of Nature. Poets of *Rig-Veda* presented *Agni*, *Vanja* and *Suria* as the highest ancient Indian Trinity. Within the *Hindu* through *Brahna*, *Vishnu* and *Shiva*, the Indian trinity components are presented not as independent, but rather as three manifestations of one and the highest divinity. This is similar to Fuller's synergy concept based on the triangle and the tetrahedron. From these basic elements, it is possible to make all other Platonic solids, including the icosahedron. There are several ways to do this. In ancient India, people took one route which led to *Five* through *four-fold* symmetry. They used both a two-dimensional method to present five elements and also a three-dimensional method to present fivefold symmetry via fourfold symmetry. The "escape" from two dimensions to three dimensions and five-fold symmetry in Hindu culture was an **orthogonal five** which allowed inclusion not only of a symmetry concept of *space*, but of *space-time* as one. For example, of the five heavenly elements *Apna*, *Prana*, *Samana*, *Vyana* and *Vdana*, the first four are represented as points in a plane, while the fifth element-point was outside of the plane, elevated from the cross section of the four planar elements: a pyramid. The Ancient Egyptians believed in a kind of *space-time synergy* displayed in their pyramids.

Fuller demonstrated his approach of synergy as a transformation of two triangles into a tetrahedron. The synergy result was $1 + 1 = 4$ because "two triangles may be combined in such a manner as to create the tetrahedron, a figure volumetrically embraced by four triangles." Fuller's starting point was a *line*, a one-dimensional entity, which was enough to solve any 3-dimensional Platonic solid. For the ancient Egyptians, a line was not enough because they were **obsessed** with solving both *3-D space* and *time*. The reason for this may be an ancient Egyptian legend:

ORTHOGONAL FIVE
                              space-time symmetry

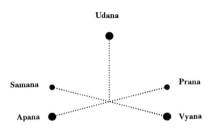

Representation of five elements in ancient India with point notation. Real link between elements is inivisible.

Egyptian pyramids as stone monuments of "petrified space-time concept" of ancient Egyptian thought.

F:2-4   **SKY-GODDESS**
                    Ancient Egyptian Temporal Icosahedron

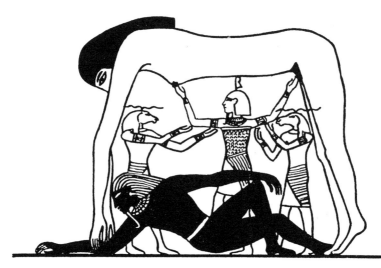

Ancient Egyptian sky-goddess Nut (light principle) and earth-god Geb (dark principle). Shu, helped
by wind spirits, separates Nut and Geb but she touched his hands and had committed a violation of the
Holy World.

In modern version (picture from 1603), sky-
goddess in sky constellation Andromeda
became despaired when she found out that her
children became sworn enemies.   She is
crucified between good and bad, but with
strong link to Holy World (belt around the
waist).

In the **World Ocean Nu** there was a *Neb-er-cher*, the *God Kepri*--meaning the **Creator**. This god is also known as **Ra**, the *Sun God* and his wife was the sky-goddess **Nut**. But he perceived that she had been unfaithful to him and he declared with a curse that she should be delivered of the child in no month and no year. In that time, the ancient Egyptian year had 360 days and 12 months, and each month had exactly 30 days. But the goddess Nut had not only the earth-god Geb as a lover, but also the god Thoth. For her love, god Thoth, playing at draught-board with the **moon**, won from him a 1/72 part of every day, which resulted in 5 additional days ($360 \div 72 = 5$). Thoth added these five new days to the Egyptian year, and since that time the year has had 365 days. On these five days, regarded as outside the year of 12 months and 360 days--that is, as epagomental--the curse of the **god-Ra** did not rest. Goddess **Nut** took that opportunity and with earth-god Geb had **five** children, one each day: three gods, *Osiris*, elder *Horus*, and *Set*; and two goddesses, *Isis* and *Nephthys*.

That was not all. According to legend, Osiris married Isis, while Set married Nephthys. But because the goddess Nut had committed a violation in the Holy World, her children were sworn enemies. Osiris was very popular (being the successor of light) and his brother Set was jealous (the successor of darkness). Set and seventy-two others plotted against Osiris. The evil brother, Set, measured Osiris' body by stealth and fashioned and decorated a coffin of the same size as Osiris' body. During a party, when all were drinking and making merry, Set brought in the coffin and jestingly promised to give it to the one whom it should fit exactly. Thinking this to be a joke, each tried it, one after the other, but it fit none of them. Last of all, the 28-year-old Osiris stepped into it and lay down, whereupon the seventy-two conspirators ran and slammed the lid down on him, nailed it fast, soldered it with molten lead, and flung the coffin into the Nile. The coffin containing the body of Osiris floated down the river and out to sea, till at last it drifted ashore at Byblus, on the coast of Syria. Isis tried to help him, but Set found the coffin as he was hunting a boar one night by the light of a full moon. He cut Osiris' body into **fourteen** pieces and scattered them abroad. Isis found the corpse of her husband, Osiris, and she and her sister, Nephthys, sat down beside it and uttered a lament which in later ages became the prototype of all Egyptian

F:2-5   ICOSAHEDRAL VIBRATIONS FREQUENCIES
                          In ancient Egypt and Fullerene Science

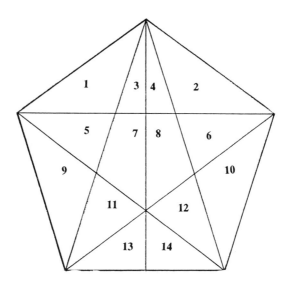

According to ancient Egyptian legend, Osiris' body was separate in fourteen pieces by Set, and Osiris died. When Isis collected the separate pieces into one part, she "fanned the cold clay with her wings" (with golden mean properties), and Osiris revived in inverse, invisible world.   Today we know that icosahedral Fullerene C_{60} has *fourteen* active frequency modes [Cm^{-1}]:   273, 437, 496, 714, 774, 1099, 1250, 1428, 1470, 1575 (Raman domain), 527, 577, 1183 and 1428 (infrared domain). Symmetry group order of icosahedron is 60, but through center of symmetry group order of C_{60} is also 120. New set of 60 symmetry elements are inversion of previous one, and for us "invisible." If fourteen frequencies coupled together, by synergy--like "golden attractor"--they may give a new 60 inversion elements as one. In both cases, ancient and modern, we have way of unity of repulsion-attraction actions of icosahedron.

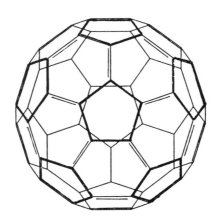

lamentations for the dead. Then Isis fanned the cold clay with her wings. Osiris revived, and thenceforth reigned as **king** over the **dead** in the *other world* of the pure light, as an invisible Universe.

Fuller, in his book, **Synergetics**, wrote that "*Negative Universe is the complementary but invisible Universe . . . The star tetrahedron may explain the whole negative phase of energetic Universe.*" According to Fuller, the invisible Universe is not some kind of fictional world, but a real one. This is similar to the concept that we have in physics of matter and antimatter. Each particle of what we call "normal matter" (e.g. electron, proton, neutron, etc.) has a complementary antiparticle (positron, antiproton, antineuron) with **mirror** properties of **charge** and **spin**. For example, the interaction of an electron and a positron will produce two **photons** as pure light. As we know, our Universe consists of matter, but in laboratories scientists have produced antimatter with very short lifetimes. Why is our Universe, according to our current knowledge, asymmetrical? Is it a normal or a "pathological" state of the Universe? What is the position of human beings in such a Universe? There are different opinions on this question in science. Nikola Tesla (1856-1943), the famous Serbian - USA inventor, pondered the question of the position of modern man in the Universe with the practical mind of an engineer: "What has the future in store for this strange being, born of a breath, or perishable tissue, yet immortal, with his powers fearful and divine? What is to be his greatest deed, his crowning achievement?" His answer was "To create and to annihilate material substance, cause it to aggregate in forms according to his desire, would be the supreme manifestation of the power of Man's mind, his most complete triumph over the physical world, his crowning achievement, which would place him beside his Creator, make him fulfill his ultimate destiny." Tesla invented *alternating current* and the *rotating magnetic field* on which modern industry is based. Tesla experienced **mind activity** at the moment when he discovered this invention, similar to the relationship between the **real world** and the **complementary one** ("world of ideas" or "imaginary world"). "Back in the deep of the brain was the solution," said Tesla, "but I could not yet give it outward expression. One afternooon in February 1882, which is everpresent in my recollection, I was enjoying a walk with my friend in the City Park and reciting poetry. At that age I knew entire books by heart, word for word. One of these was Goethe's **Faust**. The sun was just setting and reminded me of the glorious passage:

Sie rückt und weicht, der Tag ist überlebt,
Dort eilt sie hin und fördert heues Leben.
O dass kein Flügel mich vom Boden hebt,
Ihr nach und immer nach zu streben!

Ein schöner Traum, indessen sie entweicht.
Ach! zu des Geistes Flügel n wird so leicht
Kein köperlicher Flügel sich gesellen.

[The sun retreats–the day, outlived, is o'er
It hastens hence and lo! a new world is alive!
Oh, that from earth no wing can lift me up to soar
And after, ever after is to strive!

A lovely dream, the while the glory fades from sight.
Alas! To wings that lift the spirit light
No earthly wing will ever be a fellow.]

As I uttered these inspiring words, the idea came like a *flash of lightning* and in an instant the truth was revealed. I drew with a stick on the sand the diagrams shown six years later in my address before the American Institute of Electrical Engineering, and my companion understood them perfectly. The **images** I saw were wonderfully sharp and clear and had the solidity of metal and stone, so much so that I told him: See my motor here; watch me reverse it. I cannot begin to describe my emotions. Pygmalion seeing his statue come to life could not have been more deeply moved. A thousand secrets of nature which I might have stumbled upon accidentally I would have given for that one which I had wrested from her against all odds and at the peril of my existence."

From the synergy point of view, according to Fuller and Tesla, it is possible to represent an invisible Universe using tetrahedra (or discovering a **new** kind of device as a result of unity of real world (brain) and complementary one (mind). Fuller did not tell us how to do it, nor what the dimensionality of this "negative" Universe is, as complementary one to the 3-dimensional one.

The facts we know today are that (1) Goethe's *Faust* was written by the *Golden mean* law, discovered by Rakocevic, (2) Tesla was reciting *Faust* when he tried to find solution for his motor, (3) there are biological structures in the cell which play an information role and possess properties of the golden mean. Is the golden mean law a link between the real world (i.e. the 3-dimensional world and our consciousness state) and the complementary one--whose dimension we do not yet know (our subconscious)? If so, what is the dimensionality of the world complementary to our 3-D world? Independent of the concepts of the ancient Egyptians (the "other world" as an invisible Universe), modern physics (antimatter) and Fuller (negative complementary Universe), we will consider here another possible general approach from information theory.   Koruga (1991) considered neurocomputing from a geometrical-topological approach and found a principle of **information complementarity** based on an inversion operation of symmetry which is, in some ways, similar to the previous ideas.

If we use a Euclidean space, it is possible to write the Pythagorean distance in quadratic form:

$$X_1^2 + X_2^2 + X_3^2 \ldots + X_n^2 = r^2 \qquad (2.1.1)$$

which represents a sphere of radius r. It is well known that the formula for the volume of a sphere is

$$V_n(r) = C_n r^n \qquad (2.1.2)$$

where C_n is a constant (unit sphere):

$$C_n = \frac{2\pi}{n} C_{n-2} \qquad (2.1.3)$$

Now we can compute the volume of the unit sphere in N-dimensions. For the sake of simplicity, let's start with N=3. To calculate the 3-dimensional unit sphere, we should use Equation (2.1.3):

F:2-6

Table 1.1.1   The value of the unit sphere for N=2 to N=n

N	$\mathbb{C}_n$	VALUE
2	π	3.141
3	$4\pi/3$	4.188
4	$\pi^2/2$	4.934
5	$8\pi^2/15$	5.263
6	$\pi^3/6$	5.167
7	$16\pi^3/105$	4.724
.		
.		
.		
n	$\pi^n/n\ !$	$\to 0$

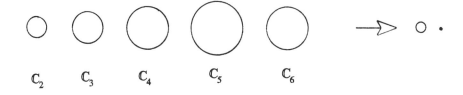

$\mathbb{C}_2$ $\mathbb{C}_3$ $\mathbb{C}_4$ $\mathbb{C}_5$ $\mathbb{C}_6$

If any of the unit spheres contain in side of self dimension N=1 (which unit sphere has value 2, black-white, but in macro-dimension is line) unit sphere or sphere of that unit sphere, will be filled by dark halo, if dimension N=1 vibrates as one-dimension object (string).

$$C_3 = \frac{2\pi}{n} \cdot C_1.$$

As we know from equation (2.1.2), the sphere volume for $N=3$ is

$$V_3(r) = \frac{4\pi}{3} r^3 , \textit{ and } \quad \frac{4\pi}{3} = \frac{2\pi}{3} \cdot C_1$$

which means that C_1 **must be** 2. This is a surprising result, but crucial to understanding: (1) Fuller's concept of synergy based on the tetrahedron, (2) ancient Egypt, and (3) this book. Why is this so important?! It shows that a *unit sphere* of a one-dimensional world, or line, has value 2. This is exactly what Fuller used for his synergy concept from line through unit triangle to tetrahedron, associated with *left* helix and *right* helix. A unit sphere of line has value 2 as its left and right rotation. In biology, we also know that proteins, as basic elements of biological life, consist exclusively of **left** amino acids, which indicates from a synergy point of view that have to exist structures with right orientation, or we are incomplete, a partially degenerative system. Can Platonic solids be made from only the left parts of a tetrahedron!?

For $N=4$ we can write

$$C_4 = \frac{2\pi}{n} C_{n-2} = \frac{2\pi}{4} C_{4-2} = \frac{2\pi}{4} \cdot C_2 = \frac{2\pi}{4} \cdot \pi = \frac{\pi^2}{2}.$$

Now it is easy to calculate the value of C_n for N = 5, 6, 7 . . . n, and we have the solution given in Figure 2-6. We can see that the volume of the unit sphere is **maximum** for N=5, and it goes to zero for N → n. This surprising result implies that the **maximum** storage of *mass, energy* and/or *information* is in a 5-dimensional unit sphere, as some kind of **super-icosahedron**.

The final step is to calculate the value of the unit sphere for dimensions 0, -1, -2, . . ., -n, or what Fuller called "negative, complementary but invisible Universe," (or the ancient Egyptians' other-invisible world). In accordance with modern physics, we use the term "world of shadow" for the Universe of unit spheres with dimensions . . . ., -2 , -1. Results based on Equation (2.1.3) are given in Figure 2-7. Interesting findings include the value of the unit sphere for dimension equal **zero** being equal to *one*. This is **unity** of two

UNIT SPHERES, DIMENSIONS AND DIMENSIONALITY
Symmetry of information complementarity

$$N = 6 \qquad \mathbb{C}_6 = \frac{2\pi}{n}\mathbb{C}_4 = \frac{2\pi}{6}\cdot\frac{\pi^2}{2} \qquad = \frac{\pi^3}{6} \qquad = 5.1677$$

$$N = 5 \qquad \mathbb{C}_5 = \frac{2\pi}{n}\mathbb{C}_3 = \frac{2\pi}{5}\cdot\frac{4\pi}{3} \qquad = \frac{8}{15}\pi^2 \qquad = 5.2637$$

$$N = 4 \qquad \mathbb{C}_4 = \frac{2\pi}{n}\mathbb{C}_2 = \frac{2\pi}{4}\cdot\pi \qquad = \frac{\pi^2}{2} \qquad = 4.9348$$

$$N = 3 \qquad \mathbb{C}_3 = \frac{2\pi}{n}\mathbb{C}_1 = \frac{2\pi}{3}\cdot 2 \qquad = \frac{4\pi}{3} \qquad = 4.1887$$

$$N = 2 \qquad \mathbb{C}_2 = \frac{2\pi}{n}\mathbb{C}_0 = \frac{2\pi}{2}\cdot 1 \qquad = \pi \qquad = 3.1415$$

$$N = 1 \qquad \mathbb{C}_1 = \frac{2\pi}{n}\mathbb{C}_{-1} = \frac{2\pi}{1}\cdot\frac{1}{\pi} \qquad = 2 \qquad = 2$$

$$N = 0 \qquad \mathbb{C}_0 = \frac{2\pi}{n}\mathbb{C}_{-2} = \frac{2\pi}{\frac{3}{2}}\cdot\frac{1}{\frac{4\pi}{3}} \qquad = 1 \qquad = 1$$

$$N = \overline{1} \qquad \mathbb{C}_{-1} = \frac{2\pi}{n}\mathbb{C}_{-3} = \frac{2\pi}{4}\cdot\frac{1}{\frac{\pi^2}{2}} \qquad = \frac{1}{\pi} \qquad = 0.3184$$

$$N = \overline{2} \qquad \mathbb{C}_{-2} = \frac{2\pi}{n}\mathbb{C}_{-4} = \frac{2\pi}{5}\cdot\frac{1}{\frac{8\pi^2}{15}} \qquad = \frac{1}{\frac{4\pi}{3}} \qquad = 0.2387$$

$$N = \overline{3} \qquad \mathbb{C}_{-3} = \frac{2\pi}{n}\mathbb{C}_{-5} = \frac{2\pi}{6}\cdot\frac{1}{\frac{\pi^3}{6}} \qquad = \frac{1}{\frac{\pi^2}{2}} \qquad = 0.203$$

$$N = \overline{4} \qquad \mathbb{C}_{-4} = \frac{2\pi}{n}\mathbb{C}_{-6} = \frac{2\pi}{7}\cdot\frac{1}{\frac{16\pi^3}{105}} \qquad = \frac{1}{\frac{8\pi^2}{15}} \qquad = 0.1899$$

F:2-8

INVERSION PROPERTY OF GOLDEN MEAN BASED ON FIBONACCI SERIES

$\Phi!/0 = \infty$	$0/\Phi! = 0$
$1/0! = 1$	$0!/1 = 1$
$2/1 = 2$	$1/2 = 0.5$
$3/2 = 1.5$	$2/3 = 0.6$
$5/3 = 1.6$	$3/5 = 0.6$
$8/5 = 1.6$	$5/8 = 0.62500$
$13/8 = 1.62500$	$8/13 = 0.61538$
$21/13 = 1.61538$	$13/21 = 0.61904$
$34/21 = 1.61904$	$21/34 = 0.61764$
$55/34 = 1.61764$	$34/55 = 0.61818$
.	.
.	.
.	.
$(\sqrt{5}+1)/2 = 1.61803 = \bigstar = 1/\bigstar + 0!$	$(\sqrt{5}-1)/2 = 0.61803 = 1/\bigstar = \bigstar - 0!$

$\bigstar = \tau = \phi$ Fibonacci Number; 0,0!,1,2,3,5,8,13,21,34,55,89 Fibonacci Series
$\phi^{0!} = \Phi!$ 0! Factorial zero (null); $\Phi!$ Njegoš scenario number (NSN) from his poem, "The Ray of Microcosmos," which was written by Golden mean.

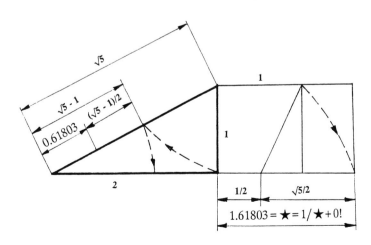

Construction of the golden mean values on triangle AB=2, BC=1 and AC=√5. Both (√5 - 1)/2 and (√5 + 1)/2 are presented.

complementary worlds, and we will call it the Light world (Lw), or invariant world. This means that the link between the real world (R_w) and shadow world (S_w) must be:

$$R_w^N = \frac{1}{S_w^{(1-N)}}$$ (2.1.4)

because, as shown in Figure 2-7, there is complementarity between R_w and S_w:

except for N=1 (dark world-Dw, because there is no complement in the shadow world and can be no light), where: N-dimension ($-\infty < N < \infty$) and n-dimensionality ($n > 1$).

$$R_w^N \cdot S_w^{(1-N)} = 1$$ (2.1.5a)

$$R_w^n \cdot S_w^{n+2} = 1$$ (2.1.5b)

Interesting is that the negative dimension N = -1 (or N = 1_4) has also properties of four-dimensionality (n=4), and we can write

$$X_1^2 + X_2^2 + X_3^2 - [(-1)_4]^2 = 0$$ (2.1.6)

which from physics we know is equivalent to Minkowski's approach to four-dimensional space-time structures, because

$$X_1^2 + X_2^2 + X_3^2 - (ct)^2 = 0$$ (2.1.7)

where: c - the speed of light and t - time.

Two things are now important. First, the fourth-dimensionality of N = -1 as n = 4, could be represented in the real world by the same law in N = 2, because $N(-1)_4 \cdot N(2) = N(0)$. Second, the linkage should be a synergetic one (in a sense; *left* and *right* helixes). The left and right parts should be structured in a plane (N = 2). Two parts from relation $N(-1)_4 \cdot N(2) = N(0)$ which can be added as left and right one can be as **logarithmic spirals** (because logarithm of b $\cdot$ c = a is *log* b + *log* c = *log* a).

Note that *N* is dimension, but *n* is dimensionality. For the positive dimensions: N = 1, 2, 3, . . ., dimension and dimensionality have the same

F:2-9   **GOLDEN SPIRALS AND GOLDEN ATTRACTOR**

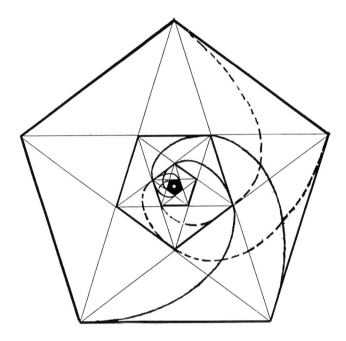

Construction of the logarithmic spiral with left and *right* orientation as property of five-fold symmetry.

Attractor with visual perception of the golden mean through golden angle of logarithmic spirals. This kind of attractor (vibro-rotational oscillator) with ortogonal property of its structure possesses spatio-temporal symmetry transformation.

F:2-10

THE ICOSAHEDRAL DESIGN OF EGYPTIANS' PYRAMID

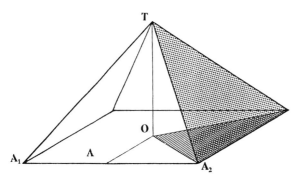

1. The projected area of a face is the golden proportion with that face.

$TO=h=146.8$ m; $A_1A_2=a=230.5$ m
$OA=C=165.25$ m
$TA=b=186.63$
$c/b=\varphi$ $\cos\alpha=\varphi$
$\alpha=\cos^{-1}0=51°49'38.3''$

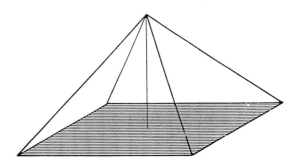

For golden mean:
$\cos\alpha=\sin^2\alpha=\cot^2\alpha=\varphi$
$\cos\alpha=\varphi=c/b$
$ca/2 : ba/2=\varphi$ (same as 1)

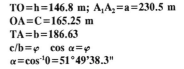

$\cot^2\alpha=\varphi$

2. The basic square, which is the projection of the pyramid, is the golden section of the visible surface.

Basic area: $E=53,130.25$ m^2
Facial area: $F=86,038.95$ m^2

$E/F = F/(E+F) = \varphi$
or
the visible area of the Pyramid is the golden section of the total area.

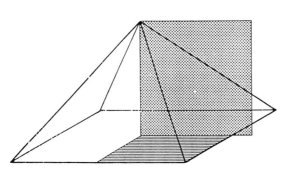

3. The square on the half base, as a fourth of the basic area, is the golden section of the square on the height.

value, while for negative dimension N = -1, -2, -3, . . ., dimension and dimensionality have relationship as N = **(3-n)**.

The logarithmic spiral is a self-similar object which we can write

$$r(\theta) = r_0 \exp(k\theta) \qquad (2.1.8)$$

where: θ is angle, $r_0 > 0$ and k are constants. Here, we are considering unit entity, like unit sphere, spiral and the value $r_0 = 1$, in the future. Since the angle θ is defined only module 2π, scaling factors equate to

$$s = \exp(2\pi mk) \qquad (2.1.9)$$

where m is an integer and s means spiral. The logarithmic spiral for k = -$(\pi/2)$ log $(\sqrt{5} - 1)/2$ gives *exactly* the property of a pentagon, and it is possible to make left and right spirals. The logarithmic spiral of the pentagon is based on the property of the **golden mean** $[(\sqrt{5} - 1)/2]$. This means that the link between real world (R_w) and shadow world (S_w) follows the law of the golden mean. If the ancient Egyptian legends made sense to those who lived in that time, then the Egyptian pyramids must be built by the **golden mean** property in N = 2 dimension. This implies that the relationship between surfaces, as a two-dimensional property, of a pyramid should have been built by the golden mean.

Let's consider the Great Pyramid, which was built between 8000 and 2600 B.C. The precise time of its construction is unknown, but many experts believe it was during the period of Cheops, about 2640 B.C. The length of the base is about 230.5 m, height 146.8 m and slope of the faces about 51° 50'. Simple calculation shows that the angle of each of the four triangular faces is a right angle which gives the projected area of a face on the base is the **golden section** of the face, and the surface of the base is also the golden section of the pyramid. There is another strange property: "*the visible area of the Pyramid is the golden section of the total area,*" according to Hugo Verheyen (1992), an expert on Egyptian pyramids, who also explained the *icosahedral design* of the Great Pyramid.

In the ancient Egyptian era, Ra, Osiris and Horus (the younger son of Osiris and Iris) were the major gods: *Sun God - Ra*, because he became one (N=5) with his *own shadow* $[N = (-4)_7]$ created **the world** and an **Ennead** or circle of *nine* basic numea, recognized by all the temples of Egypt; *Osiris,*

SUPER-ICOSAHEDRON, ORTHOGONALITY AND SCHEDOW MODEL
symmetry

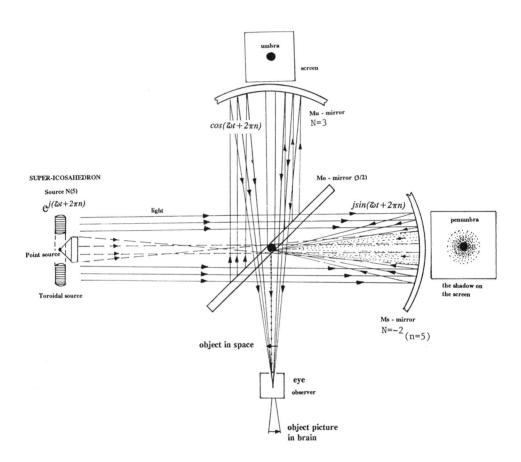

The Shadow world (Sw) model can be illustrated as a physical one. Source of light is entity of N=5, which has point source (1) and toroidal source (2). Light goes to mirror (Mo) in the center of which lies a unit sphere of dimension N=0 and value 1. Let mirror posses properties that allows it to both reflect and refract light. In this case, interaction of dimensions N=5 and N=0 will produce Shadow in mirror Ms (umbra and penumbra), while on mirror Mu interaction will produce only umbra. Mirror Ms will reflect and focus light and shadow at one point (N=0) on mirror Mo, while mirror Mu will reflect and focus light to observer. Focussing light and shadow from mirror Ms will be reflected by mirror Mo to observer. Reflecting light from mirror Mu and reflecting light Mo(Ms) will come in at the same time to observer, but shadow will be invisible to observer. This is the case whether the observer stays in one place or moves with a velocity much less than the speed of light. If observer travels with a velosity of the speed of light, the model predicts unity of observer and shadow, and world of shadow will be visible to observer. This model indicates that light is a manifestation of N=5 or super-icosahedron.

F:2-12

Dimension and dimensionality representation

Dimension	Dimensionality	Unit Sphere	Unit sphere value	Diameter exponent	Space (space-time)	Name
-2	5		$3/4\pi$	$1/r^2$		cone + polar circular cylinder
-1	4		$1/\pi$	$1/r$		polar circular cone + sphere
0	3/2		1	1		golden mean attractor
1	1		2	r^1		line
2	2		π	r^2		area
3	3		$4\pi/3$	r^3		sphere
4	4		$\pi^2/2$	r^4		spheroidal circular cone with cardinal measure $3 \cdot 10^{10}$
5	5		$8\pi^2/15$	r^5		torusoidal circular cone with cardinal measure $3 \cdot 10^{10}$

because he became the king of the *shadow world* [N = $(-1)_4$, our unconsciousness], and finally *Horus*, son of Osiris, who became the god of the **unity world** (self), *real* (N=3) and *shadow* [N = $(-2)_5$], "the sky and what is in it, the earth and all that is upon it."

In ancient notation, Osiris had the number 3, Isis had 4 and Horus had 5. What kind of synergy concept did the ancient Egyptians use for this notation? According to one simple notation: **father + mother** = *son*, should be 3 + 4 = 7. But Horus as the son has notation 5. Under which conditions does 3 + 4 = 5? One answer is $3^2 + 4^2 = 5^2$. This is the same synergy concept, area as N(2), which they used for the Great Pyramid.

As we can see, there are different concepts of synergies (Fuller, ancient Egyptians, etc.), and we are looking for one which can solve the problem of "artificial life." Our ultimate goal is to make nano-machines which will be complementary to biological molecules. We presume that this will be possible if we understand the basic principles of biomolecules as "nano-machines," and the synergetic processes in biological entities based on the golden mean, and if we find a basic complementary material to build a new kind of nano-machines based on synergy which also should be complementary to biological ones. Principles of icosahedrons and the C_{60} molecule appear to be good candidates.

But first, to go from nanobiology to nanotechnology, we should find biological molecules which possess properties of logarithmic spiral (golden mean), and also have *left* and *right* spiral orientations, to be perpendicular (synergetic condition for spatio-temporal structures). This will be done in detail in Chapter 5.

To understand *icosahedral symmetry* it is necessary to give a fundamental understanding of group theory and symmetry, and in that way to link ancient and modern approaches to the **icosahedron.**

2.2 Group Theory and Symmetry

Entities which may differ among themselves (students in a classroom, components of a car, books in a library, etc.) are called group elements, and the collection of group elements is known as a **set**. Set theory is a large subject; we consider only that part necessary to define symmetry operations. A *collection* within a set need not define the relationship among the set's entities. We need to define **group** and **set** to consider symmetry.

In any set, the elements are related to each other in some way. If we have two elements **A** and **B**, the relationship between them defines a third element **C**, or **AB**=**C**. Different relationships between **A** (e.g. A=10) and **B** (e.g. B=2) can give different values of C (10+2=12; 10x2=20; 10÷2=5, 10^2=100, etc.). Consequently, we define a group in the sense of a symmetry transformation. There are four rules which define the relationship among the elements of a set to form a group.

Closure rule: For every ordered pair of elements **A** and **B** in the set there is a unique element **C** which is also a member of the set (if AB=C, where A, B, and C are **all** elements of the set).

Associativity rule: If more elements than three are related, the resulting element must be independent of the way in which two adjacent elements are paired (A[BC]=D then [AB]C=D, so that if BC=E and AB=F, then AE=EC=D).

Identity element: According to set theory, the set must contain a unique element labelled E such that for all elements of the set EA=AE=A, which we call the identity of the unit element.

Inverse element: For every element A in the set, there must exist **in the set** a unique element labelled A^{-1} such that $AA^{-1}=A^{-1}A=E$ (if there is AB=BA=E, then $B=A^{-1}$ and $A=B^{-1}$).

If there are relationships among elements in a set such that these four rules are fulfilled, then the set of elements is called a **group**. The *order* of the group is defined by the number of elements in the group.

Usually AB≠BA, which is to say that A and B do not commute. Using the definition of a group, we can find with which group elements they commute. By definition of a group, the **identity** element (E) and **inverse** elements (A^{-1}, B^{-1}) commute with all elements of the group. In mathematics, a group in which every element commutes with every other element is called an Abelian group.

Considering that elements usually do not commute, we should define processes which lead to commutation like pre-multiplication, post-multiplication and multiplication.

In the **pre-multiplication** process, we pre-multiply both sides of the equation by the same element (AB=C then DAB=DC. If D is inverse element A^{-1} then $A^{-1}AB=A^{-1}C$ which gives $B=A^{-1}C$, because $AA^{-1}=E$).

In the **post-multiplication** process, we post-multiply both sides of the equation by some element D (ABD=CD and if $D=B^{-1}$, then $ABB^{-1}=CB^{-1}$ or $A=CB^{-1}$.).

SYMMETRY OPERATIONS
Examples

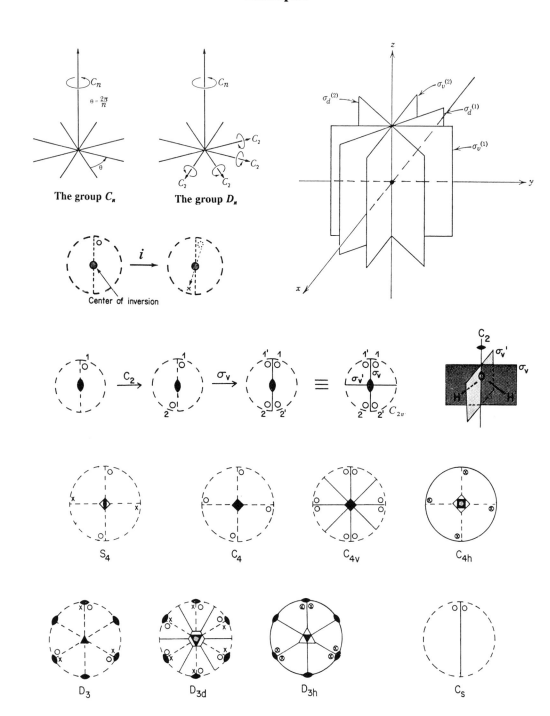

The group C_n

The group D_n

Center of inversion

C_{2v}

S_4

C_4

C_{4v}

C_{4h}

D_3

D_{3d}

D_{3h}

C_s

In the **multiplication** process, we use inverse elements to multiply AB=CB with B^{-1}, which gives A=C. But we can also divide by B. In this case, AB=BC becomes A=BCB^{-1} or C=B^{-1}AB. Multiplication data can be systematized in a multiplication table which requires knowledge of the relationship among the elements and allows us to find C for any relationship AB in AB=C.

If some elements of the set obeys the four rules for a group, it is a subset, which itself is a group and is called a **subgroup**.

Groups can be constructed by successive multiplication of any elements. If a group has A, B, C . . . elements, the first multiplication could be CC=B, the second CCC=CB=G, the third CCCC=J . . ., etc. In this case, we can write C^n=E and the group is known as a **cyclic group**.

If elements are related to one another as $S^{-1}AS$=B, the relationship is called a **similarity transformation**, where A is transformed into B by similar element S to A. If A, B, and S are all elements of the same group, B can be **conjugate** to A [$S^{-1}AS$=B gives AS=SB and A=SBS^{-1}=$(S^{-1})^{-1}BS^{-1}$]. It is possible to show that if B is conjugate to A, and C is conjugate to A, then B is conjugate to C. This operation gives a concept of a **class** which is defined as a subset of the group, consisting of all the elements of the group which are conjugate to each other. Classes are mutually exclusive so that every element belongs to one and only one (onne) class. It is well known in mathematics that a group can be divided up into classes.

In an ordinary Euclidean space, there are six main symmetry operations which are defined as follows:

Rotation C_n is an operation which effects a rotation through an angle $2\pi/n$ about an axis, fixed in space, where n is an integer. This integer defines the order of the axis and is called an n-fold axis. If a system has more than one axis of symmetry, then the highest value of axis is called the **principal axis**.

The axis of symmetry (n) is an axis about which rotation of a figure superimposes on itself. Any figure has a one-fold symmetry axis, because any body rotated through 360° about any direction completely superimposes on itself. The most symmetrical geometric body is a sphere.

Vertical plane σ_v is an operation of reflection in the vertical plane, fixed in space, containing the principal axis. The operation which effects a reflection in this plane is defined as σ_y. Simply σ_y is the mirror image of an object in the vertical plane.

Horizontal plane σ_h is an operation which produces a reflection in a

SYMMETRY OPERATIONS

THE GROUP D_3

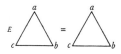

$$E^{-1}AE = A \qquad\qquad E^{-1}DE = D$$

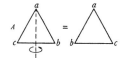

$$A^{-1}AA = A \qquad\qquad A^{-1}DA = F$$

$$B^{-1}AB = C \qquad\qquad B^{-1}DB = F$$

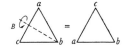

$$C^{-1}AC = B \qquad\qquad C^{-1}DC = F$$

$$D^{-1}AD = B \qquad\qquad D^{-1}DD = D$$

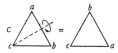

$$F^{-1}AF = C \qquad\qquad F^{-1}DF = D$$

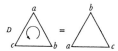

Multiplications table of the group D_3

	E	A	B	C	D	F
E	E	A	B	C	D	F
A	A	E	F	D	C	B
B	B	D	E	F	A	C
C	C	F	D	E	B	A
D	D	B	C	A	F	E
F	F	C	A	B	E	D

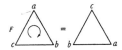

THE GROUP C_{2h}

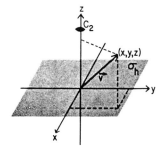

Multiplications table of the group C_{2h}

C_{2h}	E	C_2	σ_h	i
E	E	C_2	σ_h	i
C_2	C_2	E	i	σ_h
σ_h	σ_h	i	E	C_2
i	i	σ_h	C_2	E

horizontal plane perpendicular to the principal axis. The new position of the object is in the mirror image position below the plane. There is only one horizontal plane while there can be many vertical planes.

Improper rotation S_n is rotation about a principal axis fixed in space, followed by a reflection σ_h or vice versa ($S_n = C_n\sigma_h = \sigma_hC_n$).

Inversion $i = S_2$ is transformation of an object from Cartesian coordinates (x, y, z) to (-x, -y, -z). There exists a unique point, usually noted as 0, of the system such that every other point can be joined through 0 by means of a straight line to an equivalent point. 0 is generally called a **center of symmetry**.

In the case of a periodical lattice like a crystal, a set of **translational operations**, t, represents the translational displacement. The translational operation can be combined with rotation and reflection operations and give the special groups of symmetry, point symmetry groups. The C_{60} molecule is arranged by the highest point symmetry group, while its crystal state possesses one less symmetry order than does the molecule.

2.2.1 Crystal symmetry

Crystal symmetry, as one class of the point symmetry group, is well known in science. We present here only an overview to understand the crystal state of Fullerene C_{60}. We will consider the C_{60} molecule as one big atom and look at its **ordered** state within the solid, Fullerene. A crystal is composed of a regular array of atoms or, in our case, groups of sixty atoms (C_{60}), in such a way that, from a microscopic viewpoint, a section through the solid would reveal a pattern containing a periodically repeating motif. Periodicity means that *spatial* translations through distances equal to the intervals between identical sites of the atoms (molecules of C_{60}) must leave the pattern unchanged. If each atom (molecule of C_{60}) has an identical environment, that periodic array situated in space is called a **lattice**. If we can reach every point in the lattice by means of a vector translation $L = n_1\mathbf{a_1} + n_2\mathbf{a_2} + n_3\mathbf{a_3}$ where n_1, n_2 and n_3 are integers (positive or negative), then $\mathbf{a_1}$, $\mathbf{a_2}$ and $\mathbf{a_3}$ are known as basic or primitive translation vectors. These three vectors of $\mathbf{a_i}$ define an important **fundamental volume** of the space filled by the lattice which is associated with only one lattice site. The **primitive cell** is called a **fundamental volume**, and its significance is that the whole of the space filled by the lattice can be

F:2-15

CUBIC GROUPS

Type	Schoenflies symbol	A set of generators	Order	Comments
Cubic	**T**	C_3, C_2	12	The proper rotations that take a regular tetrahedron into itself; angle between C_3 and C_2 is 54.74°
	$\mathbf{T}_h$	C_3, C_2, i	24	$\mathbf{T}_h = \mathbf{T} \times \mathbf{C}_i$
	$\mathbf{T}_d$	C_3, S_4	24	All operations, including reflections, that take a regular tetrahedron into itself
	O	C_3, C_4	24	The proper rotations that take a cube or a regular octahedron into itself; angle between C_3 and C_4 is 54.74°
	$\mathbf{O}_h$	C_3, C_4, i	48	All operations, including reflections and the inversion, that take a cube or regular octahedron into itself $\mathbf{O}_h = \mathbf{O} \times \mathbf{C}_i$

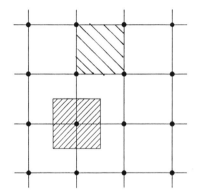

Simple cubic

Face-centered cubic

Cubic cell: upper shaded area shows a primitive cell, dots denote a lattice site, down shaded area shows cell of cubic lattice after translation.

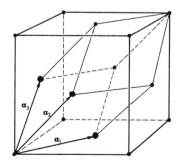

Face-centered cubic lattice as a rhombohedral primitive cell C_{60} in a solid state possesses this symmetry.

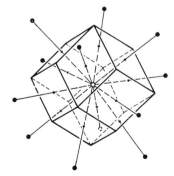

Face-centered cubic lattice after translation.

reproduced by systematically stacking the primitive cells together. A primitive cell associated with a simple cubic lattice can be chosen as a cube, while the primitive cell of a face-centered cubic lattice possesses an awkward shape.

Rotation of a crystal lattice, C_n cyclic group, gives the only possible 1, 2, 3, 4 and 6-fold rotational symmetry axes for any crystal. Based on this restriction, it is possible to define 11 more groups for crystals: D_n, T, O, S_n, C_{nh}, C_{ny}, D_{nh}, D_{nd}, T_h, T_d and O_h. The number of permitted point groups that can be associated with crystal lattices is 32. The limitation to 32 possible point groups implies that, in **three dimensions**, only 14 distinct lattices can exist. These are well known as the Bravais lattices (Bravais discovered them in 1848). The consideration of 14 Bravais lattices gives only **seven** of the 32 possible point groups which are needed to define their symmetry. These seven groups (*triclinic, monoclinic, orthorhombic, tetragonal, rhombohedral, hexagonal* and *eutic*) are called the **crystal systems**.

2.2.2 Limit groups of symmetry - Curie groups

A hundred years ago, in 1894, the married couple of Marie and Pierre Curie published a paper, "Sur la symétrie dans les phénomenes physiques, symmètrie d'un champ électrique et d'un champ magnétique," in which they described **seven** new symmetry groups based on physical properties of crystals. In the research of the geometry of crystal structure and the properties that depend on the defects of structure, crystals are considered as **discrete media**, while in the study of certain physical properties, like electrical, magnetic, thermal, elastic, etc., a crystal can be considered as a **homogeneous continuous medium**. In the latter case, M. & P. Curie included symmetry axes of an infinite order (∞) in their study. The point symmetry groups which include the infinite symmetry axes, based on physical properties of crystals, are called the **limit groups** of symmetry or **Curie groups**. A geometric figure corresponding to the ∞ point group is the rotating cone which integrates five symmetry elements, rotating axes *1, 2, 3, 4* and *6*, as one. The same cone which includes five more symmetry elements, *m, 2m, 3m, 4m* and *6m* (where *m* is a symmetry plane) is without rotation. The symmetry axis of this cone is polar. This limit symmetry group (∞m) is inherent in a homogeneous electrical field and represents direction of the lines of electric force. The limit symmetry group ∞/m integrates 13 point symmetry elements (1, 2, 3, 4, 6, 2, 4, 6, 6:m, 4:m, 3:m, 2:m, m) and represents a homogeneous magnetic field. A rotating cylinder is the characteristic of this group, in which the symmetry

F:2-16

CURIE GROUPS

Point symmetry groups	Limit symmetry groups (Curie groups)						
	∞	∞/m	∞m	∞2	∞mm	∞∞	∞∞m
1	+	+	+	+	+	+	+
2	+	+	+	+	+	+	+
3	+	+	+	+	+	+	+
4	+	+	+	+	+	+	+
6	+	+	+	+	+	+	+
$\bar{1}$		+			+		+
$\bar{4}$		+			+		+
$\bar{3}$		+			+		+
6/m		+			+		+
4/m		+			+		+
3/m		+			+		+
2/m		+			+		+
m		+	+		+		+
mm2			+		+		+
3m			+		+		+
$\bar{4}2m$			+		+		+
$\bar{3}m$			+		+		+
222				+	+	+	+
32				+	+	+	+
422				+	+	+	+
622				+	+	+	+
4mm					+		+
6mm					+		+
mmm					+		+
$\bar{6}m2$					+		+
4/mmm					+		+
6/mmm					+		+
23						+	+
432						+	+
$m\bar{3}$							+
$\bar{4}3m$							+
m3m							+
32	5	13	10	9	27	11	32

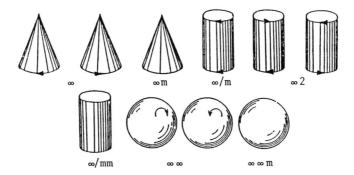

∞   ∞m   ∞/m   ∞2

∞/mm   ∞∞   ∞∞m

Relationship between Point Symmetry Groups (PSG) and Limit Symmetry Groups (LSG). Integration of 13 PSG give a ∞/m LSG (as a rotating cylinder) with magnetic properties, while 10 PSG give a ∞m LSG with electrical properties (as a cone without rotation).

plane is perpendicular to an infinite symmetry axis.

Finite geometric solids which represent Curie symmetry groups are different from Platonic ones. A cylinder at rest (∞/mm) integrates 27 point symmetry elements as a non-polar axis (∞), by an infinite number of longitudinal planes m, a simple transversal plane m, an infinite number of transversal 2-fold axes and a center of symmetry. A twisted cylinder (∞:2) integrates nine point symmetry groups, gives a non-polar axis (∞) and an infinite number of transversal 2-fold axes. Limit symmetry group $\infty\infty$ is called the group of rotation, because it is a sphere in which all its radii are rotating. This group integrates 11 point symmetry groups. Finally, the limit symmetry group which integrates all 32 point (crystal) symmetry groups is a sphere $\infty\infty$m with an infinite number of axes ∞ and an infinite number of planes, with orthogonal properties.

Three limit groups (∞, ∞2 and $\infty\infty$) are enantiomorphous, i.e., solids of this limit symmetry are right-handed or left-handed. "Using the limiting symmetry groups, Pierre Curie has established for the first time one of the most important features, distinguishing an electrical field from a magnetic one and thus managed to explain why, contrary to the case of positive and negative charges, north magnetism cannot be distinguished from south magnetism," notes Shubrikov, a world-renowned crystallographer. Shubrikov recognized how Curie's approach to symmetry is important, because ". . . it should be noted that the Curie ideas on symmetry cannot be considered as formulated in their final form. This will be made by future generations." Koruga (1992) has used Curie symmetry groups ∞/m (magnetic) and ∞m (electric) to explain structure and biophysical properties of microtubules. This will be explained further in Chapter 6.

2.2.3 Linear Infinite Order Groups (LIOG)

Linear molecules can possess one of the five symmetries: C_∞, $C_{\infty v}$, $C_{\infty h}$, D_∞ and $D_{\infty h}$. The group C_∞ consists of all rotations about a fixed axis. Because all rotations about a fixed axis commute, this symmetry group is Abelian. $C_{\infty v}$ is not an Abelian group (in spite of the fact that it is obtained from C_∞ by adding an operation σ_v as generator) because operators σ_v do not commute with rotations. All of the operations σ_v belong to the same class, since there is a rotation that will transform any vertical reflection into any other. When a horizontal plane of symmetry is added to C_∞, we have symmetry group $C_{\infty h}$. Bearing in mind that σ_h commutes with all of the rotations, this group is Abelian. When we add a C_2 axis perpendicular to the C_∞, we obtain D_∞. Because these operations do not all commute, D_∞ is not the direct product of C_∞ and C_2. But symmetry group $D_{\infty h}$ is the direct product of $D_\infty \times C_i$ and all irreducible representation can be obtained directly from those of D_∞ and C_i.

The $C_{\infty v}$ symmetry group represents the electrical field of linear molecules, while $C_{\infty h}$ represents their magnetic field. These symmetry groups are relevant to carbon clusters and their linear forms.

2.2.4 Kugel Groups: Symmetry properties of Hamiltonians, Eigenfunctions, and Atomic Orbitals

The Schrödinger equation can be written

$$H\psi = E\psi \tag{2.2.4.1}$$

where H is the Hamiltonian operator, E is a constant equal in value to the energy of the quantum state, and ψ is a wave function. The constant E is called the **eigenvalues**, while ψ is called the **eigenfunctions** of Hamiltonian. The Schrödinger equation can also be written in the form

$$\nabla^2\psi + \frac{8\pi^2 m}{h^2}(E-V)\psi = 0 \tag{2.2.4.2}$$

where ∇^2 is the Hamiltonian operator of kinetic energy. Based on this operator, in the spherical polar coordinates (r, θ, ϕ) it is possible to define the spherical harmonics $Y_{lm}(\theta, \phi)$. The symmetry group which represents the spherical harmonics is K_h (Kugel group). This group corresponds to the symmetry of the sphere, with rotation-inversion properties. Another group, K, is the subgroup of K_h which consists of all proper rotations ($K_h = K \times C_i$, where C_i is center of inversion).

The spherical harmonics $Y_{lm}(\theta, \phi)$ (m = $-l \ldots l$) are the angular parts of the eigenstates of the one-electron atom, and they are also basic functions for the irreducible representation of dimension $2l+1$ of K_h. Spectral terms, many-electron atomic functions, are also basic functions for the irreducible representations of K_h. Any unitary combination of $2l+1$ functions forms a

F:2-17

Summary of main symmetry groups

Symmetry Group	Important Symmetry Elements	Order of Group
C_1	E	1
C_i	i	2
C_s	σ	2
C_n	C_n	n
S_n	$Sn = C_{nh}$	n
C_{ny}	C_n, σ_y	2n
C_{nh}	C_n, σ_h	2n
D_n	C_n, $\perp C_2$	2n
D_{nd}	C_n, $\perp C_2$, σ_d	4n
D_{nh}	C_n, $\perp C_2$, σ_n	4n
$C_{\infty v}$	linear molecules without center of inversion	∞
$D_{\infty h}$	linear molecules with center of inversion	∞
T_d	tetrahedral symmetry	24
T_h	tetrahedral symmetry, σ_h	24
O_h	octahedral symmetry	48
I_{532}	icosahedral symmetry	60
I_h	icosahedral symmetry	120
K	spherical symmetry (Kugel group)	∞
K_h	spherical symmetry (Kugel group)	2∞
C_{u1}	limit symmetry (Curie groups) ∞	∞
C_{u2}	limit symmetry (∞/m)	2∞
C_{u3}	limit symmetry (∞m)	2∞
C_{u4}	limit symmetry ($\infty 2$)	2∞
C_{u5}	limit symmetry (∞mm)	4∞
C_{u6}	limit symmetry ($\infty \infty$)	∞^3
C_{u7}	limit symmetry ($\infty \infty m$)	$2\infty^3$

representation equivalent to dimensional irreducible representation and gives a set of eigenfunctions. The atomic orbitals are combinations that give real functions. For example, the s orbital is the totally symmetric one-dimensional representation, the p orbitals give a three-dimensional representation, d orbitals are five-dimensional representation.

The K_h group is very well defined and can be found in any textbook about atomic symmetry and its application in physics and chemistry. Another notation used in the literature for this symmetry group is $\mathbf{R_3}$.

2.3 Icosahedral symmetry

Five-fold symmetry was ignored for a long period in science, especially in crystallography. This was because of a proof of the impossibility of a 5-fold symmetry axis in a crystal medium. In crystal forms only 1, 2, 3, 4 and 6-fold symmetry axes are possible. Neither 5-fold axes nor those higher than 6-fold are possible. Three-dimensional structures with fivefold symmetry cannot be packed to fill all available space and cover a surface without gaps.

But in 1984, Schectman, Blech, Gratias and Cahn discovered quasi-icosahedral point symmetry in x-ray diffraction from the alloy Al_6Mn. This icosahedral *quasilattice* has a layer structure in E^6 as a six-dimensional Euclidean space or a hypercube. When we decompose $E^6 \rightarrow E_1^3 + E_2^3$, where $E_1^3 \perp E_2^3$, we reach rhombohedra with four edges perpendicular to a fixed face of the dodecahedron from six systems of infinite parallel layers in the quasilattice. This construction yields in a straightforward way the main global properties of the three-dimensional Penrose tilings, as quasiperiodically icosahedral symmetry in 3-D.

However, what is impossible in inorganic structures is natural in living organisms as higher dimensional entities ($N > 3$). Plants with a five-fold symmetry occur very frequently, and fivefold symmetry is also observed in the animal world. But if the golden mean is the main property of the pentagon, then principles of five-fold symmetry should be common in nature. The golden mean is a result of the Fibonacci series, which Leonardo of Pisa-Fibonacci (1180-1250) used as the solution to the famous problem concerning the propagation of populations of rabbits through successive generations.

DNA (deoxyribonucleic acids), the essential genetic structure of all living organisms, possesses fivefold symmetry properties, because the ratio of one of the periodic lengths of the ten bases of DNA to the diameter of the star decagon measures 34:20, which is the ninth member of the Fibonacci series

F:2-18

Fivefold symmetry axes and icosahedron

Numbers	Group Symbol	Group Order	Elements of Complete Symmetry
1	5	5	(1)5
2	$\bar{5}$	10	(1)5, (1)$\bar{10}$, C
3	$\bar{10}$	10	(1)5, (1)$\bar{5}$, m
4	5m	10	(1)5, (5)m
5	52	10	(1)5, (5)2
6	$\bar{10}$m2	20	(1)5,(1)$\bar{5}$, (5)m, (5)2, m
7	$\bar{5}$m	20	(1)5, (1)$\bar{10}$, (5)m, (5)2, C
8	$\underline{5}$	10	(1)5, (1)$\underline{\bar{10}}$, $\underline{C}$
9	$\bar{10}$	10	(1)5, (1)$\underline{\bar{5}}$, $\underline{m}$
10	$\underline{5m}$	10	(1)5, (5)$\underline{m}$
11	$5\underline{2}$	10	(1)5, (5)$\underline{2}$
12	$\bar{10}$m$\underline{2}$	20	(1)5,(1)$\bar{5}$, (5)$\underline{2}$, (5)$\underline{m}$, m
13	$\bar{10}$m$\underline{2}$	20	(1)5,(1)$\underline{\bar{5}}$, (5)$\underline{2}$, (5)m, $\underline{m}$
14	$\bar{10}$m$\underline{2}$	20	(1)5, (1)$\underline{\bar{5}}$, (5)2, (5)$\underline{m}$, m
15	$\underline{\bar{5}m}$	20	(1)5, (1)$\bar{10}$, (5)$\underline{2}$, (5)$\underline{m}$, C
16	$\underline{\bar{5}m}$	20	(1)5, (1)$\underline{\bar{10}}$, (5)$\underline{2}$, (5)m, $\underline{C}$
17	$\underline{\bar{5}m}$	20	(1)5, (1)$\underline{\bar{10}}$, (5)2, (5)m, $\underline{C}$
18	$\underline{10}$/$\underline{mmm}$	40	(1)$\underline{10}$, (1)$\bar{10}$, (1)5, (1)$\underline{\bar{5}}$, (5)$\underline{2}$, (5)2, (5)m, (5)$\underline{m}$, $\underline{2}$, $\underline{m}$, C
19	$\underline{10}$/$\underline{mmm}$	40	(1)$\underline{10}$, (1)$\bar{10}$, (1)5, (1)$\underline{5}$, (5)$\underline{2}$, (5)2, (5)m, (5)$\underline{m}$, $\underline{2}$, m, $\underline{C}$
20	$\underline{10}$/m	20	(1)$\underline{10}$, (1)$\bar{10}$, (1)5, (1)$\underline{5}$, (1)$\underline{2}$, (1)m, $\underline{C}$
21	$\underline{10}$/m	20	(1)$\underline{10}$, (1)$\bar{10}$, (1)5, (1)$\underline{5}$, (1)$\underline{2}$, (1)$\underline{m}$, C
22	10m$\underline{m}$	20	(1)$\underline{10}$, (1)5, (5)m, (5)m, (1)2
23	$\bar{10}$2$\underline{2}$	20	(1)$\underline{10}$, (1)5, (5)$\underline{2}$, (1)2, (1)$\underline{2}$
24	$\underline{10}$	10	(1)$\underline{10}$, (1)5, (1)2
25	$\bar{10}$3m	120	(6)5, (6)$\bar{5}$, (10)3, (10)$\bar{3}$, (15)2, (15)m, C
26	$\bar{10}$3$\underline{m}$	120	(6)5, (6)$\underline{\bar{5}}$, (10)3, (10) $\bar{3}$, (15)2, (15)$\underline{m}$, $\underline{C}$
27	532	60	(6)5, 10(3), 15(2)

$\bar{5}$ - operation of inversion; $\underline{5}$ - mirror reflection, (5) - the number of the corresponding complete symmetry elements

F:2-19   **ICOSAHEDRAL SYMMETRY**

Icosahedron

20 faces (equilateral triangles)
12 vertices
30 edges

5-fold   3-fold   2-fold
axis   axis   axis

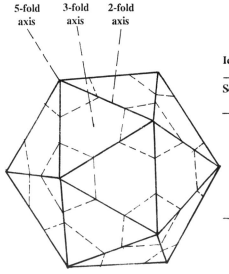

Truncated icosahedron

Icosahedral symmetries

Schoenfies symbol	A set of generators	order	comments
I	C_3, C_5	60	The proper rotations that take a regular isocahedron or dodecahedron into itself; angle between C_3 and C_5 is 37.38°
I_h	C_3, C_5, i	120	All the operations, including reflections and inversion, that take a regular isocahedron or dodecahedron into itself $I_h = I \times C_i$

120 operations:
E, $12C_5$, $12C^2_5$, $20C_3$, $15C_2$, i, $12S_{10}$, $12S^3_{10}$, $20S_6$, 15σ

C_{60} molecule
    60 carbon atoms
    12 pentagons
    20 hexagons
    30 double bonds

Angle between a double bond and its adjacent pentagonal face:

$$\psi = \cos^{-1}[(1/2)\cos(3/5)\pi]$$

Angle between two adjacent hexagonal faces is

$$\phi = \cos^{-1}[(8/3)(\sin(3/10)\pi)^2 - 1]$$

Angle between adjacent hexagonal and pentagonal faces is

$$\theta = 1/2(\pi - \phi) - \psi$$

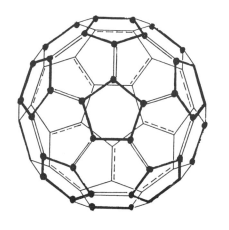

THE ICOSAHEDRAL GROUPS AND GOLDEN MEAN

Multiplication Table of the Groups I and I_h

I/I_h	E	$12C_5$	$12C_5^2$	$20C_3$	$15C_2$	i	$12S_{10}$	$12S_{10}^3$	$20S_6$	15σ	III	IV
A_g	1	1	1	1	1	1	1	1	1	1		$x^2+y^2+z^2$
T_{1g}	3	$\frac{1}{2}(1+\sqrt{5})$	$\frac{1}{2}(1-\sqrt{5})$	0	-1	3	$\frac{1}{2}(1-\sqrt{5})$	$\frac{1}{2}(1+\sqrt{5})$	0	-1	(R_x, R_y, R_z)	
T_{2g}	3	$\frac{1}{2}(1-\sqrt{5})$	$\frac{1}{2}(1+\sqrt{5})$	0	-1	3	$\frac{1}{2}(1+\sqrt{5})$	$\frac{1}{2}(1-\sqrt{5})$	0	-1		
G_g	4	-1	-1	1	0	4	-1	-1	1	0		
H_g	5	0	0	-1	1	5	0	0	-1	1		$(2z^2-x^2-y^2,$ $x^2-y^2,$ $xy, yz, zx)$
A_u	1	1	1	1	1	-1	-1	-1	-1	-1		
T_{1u}	3	$\frac{1}{2}(1+\sqrt{5})$	$\frac{1}{2}(1-\sqrt{5})$	0	-1	-3	$-\frac{1}{2}(1-\sqrt{5})$	$-\frac{1}{2}(1+\sqrt{5})$	0	1	(x,y,z)	
T_{2u}	3	$\frac{1}{2}(1-\sqrt{5})$	$\frac{1}{2}(1+\sqrt{5})$	0	-1	-3	$-\frac{1}{2}(1+\sqrt{5})$	$-\frac{1}{2}(1-\sqrt{5})$	0	1		$(x^3, y^3, z^3),$ $[x(z^2-y^2), y(z^2-x^2),$ $z(x^2-y^2), xyz]$
G_u	4	-1	-1	1	0	-4	1	1	-1	0		
H_u	5	0	0	-1	1	-5	0	0	1	-1		

A   - identity irrep

T_1, T_2 - two three-dimensional irreps

G   - four-dimensional irrep

H   - five-dimensional irrep

g   - even parity of orbitals or state (gerade)

u   - odd parity of orbitals or state (ungerade)

E, C_5, C_5^2, C_3 . . . symmetry operations

Area of symmetry operators and irreps are the characters of the irreducible representations of the group

In area III there are always six symbols x, y, z, R_x, R_y, R_z. The first three represent the coordinates, while the second three represent rotations about the axes specified in the subscripts (infrared active modes)

In area IV are listed all of the squares and binary products of coordinates according to their transformation properties (Raman active modes)

For the pure rotation group I, the outlined section in the upper left is the character table; the g subscripts should, of course, be dropped and (x, y, z) assigned to the T_1 representation. Values $1/2(1+\sqrt{5})$ and $1/2(1-\sqrt{5})$ are golden mean.

$1 \rightarrow 2$
$2 \rightarrow 3$
$3 \rightarrow 4$
$4 \rightarrow 5$
$5 \rightarrow 1$

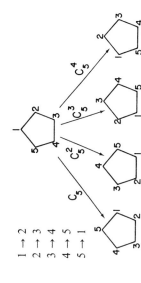

A scheme for generating reducible representations of the operations C_5, C_5^2, and C_5^3.

(golden mean) with an experimental measurement error of 0.9%. As is well known, DNA (based on **right-handed** nucleic acids) works through RNA (ribonucleic acid) to produce proteins which are based on **left-handed** amino acids. One such protein, **tubulin**, makes microtubules with **exact** properties of fivefold symmetry as a golden mean.

Recent results in physics and chemistry, such as the discovery of the C_{60} molecule, have also shown exact properties of five-fold symmetry and the golden mean in a truncated icosahedron. An icosahedron has twelve vertices, twenty faces and thirty edges by the Euler polyhedra classification. There are two groups: I (rotation) and I_h ($I \times C_i$). The first one has a symmetry order of 60, while the second has a symmetry order of 120.

The C_{60} molecule possesses symmetry properties of a truncated icosahedron. This kind of icosahedron is created when we cut vertices on approximately 1/3 of the length of an edge. In other words, we can say that each carbon atom of a C_{60} molecule in an equilibrium state is located 1/3 of the distance along the edge from each vertice.

The relationship between symmetry operations and energy state (like Hückel Molecular Orbitals - HMO) is given in a multiplication table. Analysis of the molecular orbitals are distributed into the irreducible representations of the I_h group according to:

$$\Gamma_\pi = A_g(1) + T_{1g}(3) + T_{2g}(3) + 2G_g(4) + 3H_g(5)$$
$$+ 2T_{1u}(3) + 2T_{2u}(3) + 2G_u(4) + 2H_u(5) \qquad (2.3.1)$$

where: (1), (3), (4) and (5) represent irreps dimensionality. Based on this multiplication table, it is possible to write Hückel determinants and calculate energy levels for C_{60}. These will be done in later chapters on Fullerene symmetry and electronic structures.

CHAPTER 3

SYMMETRY OF ELECTRONIC AND VIBRATIONAL STRUCTURES OF C_{60}

3.1 Carbon Atom Electronic Structure

In quantum mechanics, each wavefunction, $\psi(r, \nu, \phi)$, which is a solution of the Schrödinger equation (1.2.4.2) is known as an atomic orbital. Each electron in each orbital has a particular set of values for the four quantum numbers: n - **principal**, l - **azimuthal**, m_l -   **magnetic**, m_s - **magnetic number of electronic spin**.

The **principal** quantum number (n) gives us information about the atomic shell and its orbital energy

$$E_n = -\frac{m_e e^4}{8\epsilon_0^2 h^2} \cdot \frac{Z^2}{n^2} \qquad (3.1.1)$$

where: m_e - mass of electron, e - electron charge, ϵ_0 - permittivity of free space, h - Planck constant, Z - atomic number, and n - principal quantum number.

As we can see, the energy values are negative in sign. This occurs because the energy of completely separate protons and electrons is taken as zero. As they come together, the energy of the system is lowered below zero because of their attraction. In this case, energy is needed to separate them and bring the energy back to zero. Spacing of the levels becomes increasingly close as the value of the quantum number increases. The value of the principal quantum number is 1, 2, 3, ... For the carbon atom, n has two values, 1 and 2, because two electrons are in the first shell (**K**), while the other four are in the second shell (**L**).

The **azimuthal** quantum number (l) gives us information about the shape of the orbital and the orbital angular momentum of the electron. This quantum number has value 0, 1, 2, 3, . . ., $n-1$. Based on Schrödinger's solution, the values of the total orbital angular momentum is

$$L = \sqrt{l(l+1)} \cdot (h/2\pi) \qquad (3.1.2)$$

This shows that the first carbon orbital ($n=1$, $l=0$) does not possess orbital angular momentum. The first part of the second orbital ($n=2$, $l=0$) does not have orbital momentum either, while the second part of the second orbital ($n=2$, $l=1$) does. This means that the total angular momentum of the

THE GROUND STATE OF ATOMS
The ground state electron structures of the first eight elements in the periodic table

Element	1s	2s	$2p_x$	$2p_y$	$2p_z$	Overall
H	↑					1s
He	↑↓					$1s^2$
Li	↑↓	↑				$1s^2 2s$
Be	↑↓	↑↓				$1s^2 2s^2$
B	↑↓	↑↓	↑			$1s^2 2s^2 2p$
C	↑↓	↑↓	↑	↑		$1s^2 2s^2 2p^2$
N	↑↓	↑↓	↑	↑	↑	$1s^2 2s^2 2p^3$
O	↑↓	↑↓	↑↓	↑	↑	$1s^2 2s^2 2p^4$

THE ELECTRONIC CONFIGURATION FOR K, L and M ATOMIC SHELLS

Shell	n	l	m_l	m_s		Number of electrons	Electronic configuration
K	1	0	0	+1/2	−1/2	2	$1s^2$
L	2	0	0	+1/2	−1/2		
		1	+1	+1/2	−1/2	8	$2s^2 p^6$
			0	+1/2	−1/2		
			−1	+1/2	−1/2		
M	3	0	0	+1/2	−1/2		
		1	+1	+1/2	−1/2		
			0	+1/2	−1/2		
			−1	+1/2	−1/2		
						18	$3s^2 p^6 d^{10}$
		2	+2	+1/2	−1/2		
			+1	+1/2	−1/2		
			0	+1/2	−1/2		
			−1	+1/2	−1/2		
			−2	+1/2	−1/2		

carbon atom arises from two electrons in the other shell.

The **magnetic** quantum number (m_l) tells us the number of orbitals with a given value of l and gives information about what happens to their energies when the symmetry of the atom is disturbed. Some of the orbitals are degenerate, and the degree of degeneracy is given by the square of the principal quantum number. The magnetic quantum number may have any value of $-l$ to $+l$. Because m_l and l are related, the magnetic quantum number also gives information about angular momentum. This number also governs the number of orbitals for each value of l, as well as their behavior in magnetic or electrical fields. Only electrons of carbon atoms which have $n=2$, $l=1$ and $m_l=1$ possess angular momentum. The number of projections of m_l on magnetic direction is $2l+1$ (sum of all $+l$, of all $-l$, and 0).

The **magnetic** number of **electronic spin** (m_s) has two values, $+1/2$ and $-1/2$, and tells us about the direction of spin projection on the magnetic field.

Wolfgang Pauli published his famous principle in 1925, and said "No two electrons in the same atom can have the same set of four quantum numbers n, l, m_l, m_s." This implies that a maximum of two electrons can exist with the same spatial wavefunction, but with opposite spins. Taking the carbon atom as an example, the ground state wavefunction starts from the hydrogen orbital, which has values 1, 0, 0 for n, l, m_e, respectively. The first electron has a spin quantum number $m_s = +1/2$, and the set [1, 0, 0, 1/2] represents the four quantum numbers.   An electron configuration with $1=0$ is notified as s configuration, with notation $1s$. If the second electron goes into the same orbital, the only other possibility is the set [1, 0, 0, -1/2] for the four quantum numbers. This is the ground state of the **helium** atom, and this configuration of elections is classified as $1s^2$ (first orbital with two electrons).

The third electron would therefore go into the next available orbital with quantum numbers [2, 0, 0, 1/2] and this electronic configuration is classified as $1s^2 2s$ (s again because $l=0$, and 2 because $n=2$); this is the **lithium** atom.

The fourth electron, according to quantum mechanics and the Pauli exclusion principle, may have only [2, 0, 0, -1/2] set of quantum numbers. This electronic configuration is $1s^2 2s^2$, and this atom is **beryllium**.

The fifth electron may have a configuration determined by the following set of quantum numbers: [2, 1, 1, 1/2]. This configuration is $1s^2 2s^2 2p$ (p because $l=1$). This is the configuration of **boron**.

The sixth electron has the [2, 1, 0, 1/2] set of quantum numbers. This means that an atom of **carbon** has two electrons in $l=1$ with the same

F:3.1-2

K_h SYMMETRY AND CARBON ELECTRONIC STRUCTURE

K_h symmetry and spherical harmonics

Orbital name	Wavefunction, $\psi(r, \theta, \phi)$
1s	$\dfrac{1}{\sqrt{\pi}}\left(\dfrac{Z}{a_0}\right)^{3/2} e^{-\rho}$
2s	$\dfrac{1}{4\sqrt{(2\pi)}}\left(\dfrac{Z}{a_0}\right)^{3/2}(2-\rho)e^{-\rho/2}$
$2p_z$	$\dfrac{1}{4\sqrt{(2\pi)}}\left(\dfrac{Z}{a_0}\right)^{3/2}\rho e^{-\rho/2}\cos\theta$
$2p_x$	$\dfrac{1}{4\sqrt{(2\pi)}}\left(\dfrac{Z}{a_0}\right)^{3/2}\rho e^{-\rho/2}\sin\theta\cos\phi$
$2p_y$	$\dfrac{1}{4\sqrt{(2\pi)}}\left(\dfrac{Z}{a_0}\right)^{3/2}\rho e^{-\rho/2}\sin\theta\sin\phi$

The quantity ρ is given by $\rho = Zr/a_0$ where Z is the atomic number.

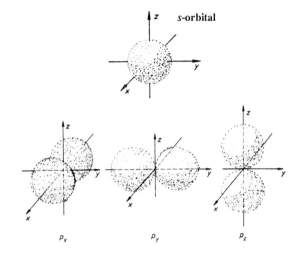

p_x p_y p_z

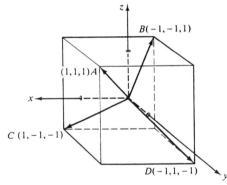

$B(-1,-1,1)$

$(1,1,1)A$

$C(1,-1,-1)$

$D(-1,1,-1)$

sp^3 - hybrid orbitals

$$\psi_1 = \tfrac{1}{2}(2s + 2p_x + 2p_y + 2p_z)$$

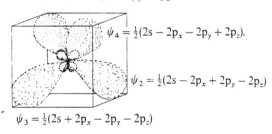

$$\psi_4 = \tfrac{1}{2}(2s - 2p_x - 2p_y + 2p_z).$$

$$\psi_2 = \tfrac{1}{2}(2s - 2p_x + 2p_y - 2p_z)$$

$$\psi_3 = \tfrac{1}{2}(2s + 2p_x - 2p_y - 2p_z)$$

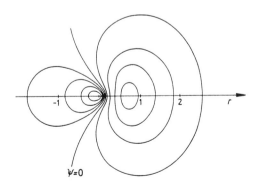

$\Psi = 0$

About 87.5% of charge is located on right side of $\Psi = 0$, while 12.5% of change is located on left side.

Orbitals and quantum numbers

Orbital	n	l	m_l	Degeneracy
1s	1	0	0	1
2s	2	0	0	
$2p_x$	2	1	± 1	
$2p_z$	2	1	0	4
$2p_y$	2	1	± 1	

F:3.1-3

DIAMOND AND GRAPHITE
First two pure forms of carbon

sp^3 - hybrid orbitals

$$\psi_1 = \tfrac{1}{2}(2s + 2p_x + 2p_y + 2p_z)$$

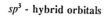
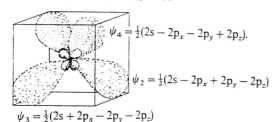

$$\psi_4 = \tfrac{1}{2}(2s - 2p_x - 2p_y + 2p_z).$$

$$\psi_2 = \tfrac{1}{2}(2s - 2p_x + 2p_y - 2p_z)$$

$$\psi_3 = \tfrac{1}{2}(2s + 2p_x - 2p_y - 2p_z)$$

Diamond

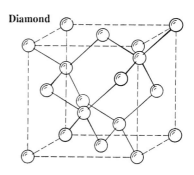

sp^2 - hybrid orbitals

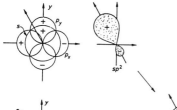
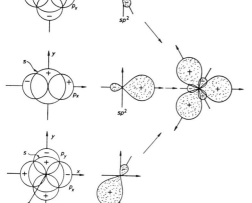

π-electrons

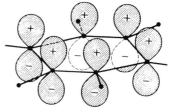

Graphite

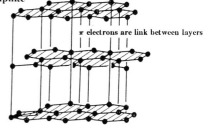

π electrons are link between layers

Five carbon atoms in unit cell of graphite

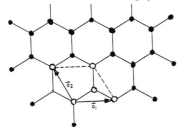

Three-dimensional unit cell of graphite

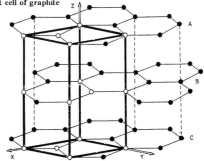

orientation of electronic spin +1/2.   This gives the ground electronic configuration of a carbon atom as $1s^2 2s^2 2p^2$.

The S-orbital is a sphere with its center in an atomic nucleus with radius r=1, while the p-orbital has radius $r=\sqrt{3} \cdot cos \vartheta$ and there are two spheres per coordinate (x, y, z) which are touching each other in the center of the nucleus.

It is well known that the basic carbon state $[He]2s^2 p_x p_y$ cannot explain its four valence bonds, while its excitation state $[He]2s p_x p_y p_z$ cannot explain its tetrahedron valence bonds.   But hybridization of one s-orbital and three p-orbitals gives carbon tetrahedron valence bonds.   Hybridization is a mathematical explanation which does not definitely exist in reality.   If hybridization does exist, it is a property of degenerating wave states, because two wave states may couple if their energies are equivalent or very close. However, there is a big difference between the energies of $2s$ and $2p$ orbitals, which argues against hybridization.   A possible explanation is the influence of atomic fields on orbitals such that s and p orbitals have very close energy values.   The building of a hybrid orbital (like sp^3) relies on the assumption of a $2s$ electron to a $2p$ orbital.   This requires that energy be put into the system, or that $2s$ electrons in many atom systems, as collective phenomena, can be excited into p electron states in carbon atoms.   Thus $2s$ electrons in carbon systems (graphite, diamond, C_{60}) can produce an excited state as a collective phenomenon of a large number of angular momenta, and forbidden transitions. In this case, conditional hybridization of s and p orbitals is possible as sp^3 and each orbital can be normalized to give:

$$\psi_1 = 1/2 \ (s + p_x + p_y + p_z)$$
$$\psi_2 = 1/2 \ (s - p_x - p_y + p_z)$$
$$\psi_3 = 1/2 \ (s + p_x - p_y - p_z)$$
$$\psi_4 = 1/2 \ (s - p_x + p_y - p_z).$$

(3.1.3)

If we use the intensity of the s orbital as 1, then the pure p-orbital has a value of 1.73, sp-orbital has a value of 1.93, the sp^2-orbital has a value of 1.99, while the sp^3 has an intensity of 2.   Indeed, the principle of maximum overlap says that the greater the overlap, the stronger the bond.

The pure carbon form, **diamond**, exists as an arrangement of the four sp^3 orbitals in tetrahedral directions.   The angle between sp^3 orbitals is 109°28'.   There is also another pure carbon form, **graphite**, with a different hybridization of orbitals. In this case, one s-orbital and two p-orbitals (p_x, p_y) give three sp^2 hybridization orbitals per carbon atom, with an angle of 120° in the plane of the σ bonds. In this case, the p_z electron of carbon atoms in graphite will form π-electrons perpendicular to the graphite plane or σ bonds. These electrons (p_z) are responsible for the link between graphite layers.

3.2 C_{60} Electronic Structure

The C_{60} molecule is the third known pure crystal form of carbon, in addition to graphite and diamond. The electronic structure is a complex, "many body" problem, because there are 60 x 6 = 360 electrons. Conversely, the C_{60} molecule has attributes of a "big atom," because it has a close spherical electronic shell and possesses unique icosahedral symmetry properties. In the truncated icosahedral structure there are two characteristic C-C bond lengths: C_5-C_5 in pentagons, C_5-C_6 double bonds in hexagons (or link between two pentagons), while in other Fullerenes there may exist a third bond, and C_6-C_6 in hexagons. There are sixty carbon p_z orbitals, each pointed along radial axes. If interactions among p_z orbitals belonging to carbon atoms on a certain pentagon are considered and interactions among orbitals located on different pentagons (there are 12 such pentagons) are neglected, then the five eigenstates based on K_h symmetry (spherical harmonics) can be written as

$$\psi_m = (\sqrt{5})^{-1} \sum \exp(\frac{2}{5}im\pi n)\phi_n \qquad (3.1.4)$$

where m = -2, -1, 0, 1 and 2 (rotational quantum numbers about the pentagon axis), and ϕ_n represents the five successive atomic orbitals on the pentagon. The corresponding eigenenergies are

$$E_m = 2\beta \cos(\frac{2}{5}m\pi) \qquad (3.1.5)$$

where β is the resonance integral. For M = 0 the energy is 2β, while for m = +1 and -1 the energy is 0.62β, and finally for m = +2 and -2 the energy is -1.62β.

There are three sets of orbitals which occur grouped together: ψ_0, $\psi_{1(+,-)}$, and $\psi_{2(+,-)}$. Interaction among 12 pentagons will split the twelve ψ_0 orbitals to $A_g + H_g + T_{1u} + T_{2u}$, while the 24 $\psi_{1(+,-)}$ orbitals, two per pentagon, will split into $T_{1g} + G_g + H_g + T_{1u} + G_u + H_u$ irreducible representations (or symmetries). The final 24 $\psi_{2(+,-)}$ orbitals, with the highest energy, will be split into T_{2g}, G_u, G_g, H_u, T_{2u} and H_g.

Based on the Hückel determinant $|A + xE|$, where A is the adjacency matrix and E the unit matrix, it is possible to calculate irreducible representations (A_g, T_{1g}, . . .) from Equation 2.3.1 and the multiplication table of the icosahedral group $[\tau = 1/2(1+5^{1/2})]$. This calculation will give Hückel energy levels for the C_{60} molecule in terms of x values based on energies in β units (β = -0.2 Rydberg = -2.72 eV) for the hypothetical planar case:

$$A_g |x+3| \rightarrow A_g(1) = -3$$
$$T_{1g} |x-\tau^2| \rightarrow T_{1g}(3) = 1/2\,(3-5^{1/2}) = 0.3819$$
$$T_{2g} |x-\tau^2| \rightarrow T_{2g}(3) = 1/2\,(3+5^{1/2}) = 2.6180$$

$$G_g \begin{vmatrix} X+1/2 & 1/2(5)^{1/2} \\ \\ 1/2(5)^{1/2} & x-3/2 \end{vmatrix} \rightarrow G_g(4) = \begin{cases} -1 \\ \\ 2 \end{cases}$$

$$H_g \begin{vmatrix} x+1/2 & 0 & 2/5(6)^{1/2} \\ 0 & x+1/2 & 1/2(5)^{1/2} \\ 2/5(6)^{1/2} & 1/2(5)^{1/2} & x-3/10 \end{vmatrix} \rightarrow H_g(5) = \begin{cases} -2.3027 \\ \\ 1.3027 \end{cases}$$

$$T_{1u} \begin{vmatrix} x+\tau^2 & -\tau^{-1} \\ \\ -\tau^{-1} & x \end{vmatrix} \rightarrow T_{1u}(3) = \begin{cases} -2.7566 \\ \\ 0.1385 \end{cases}$$

$$T_{2u} \begin{vmatrix} x & -\tau \\ \\ -\tau & x+\tau^2 \end{vmatrix} \rightarrow T_{2u}(3) = \begin{cases} -1.8202 \\ \\ 1.4383 \end{cases}$$

$$G_u \begin{vmatrix} x & 2 \\ \\ 2 & x-1 \end{vmatrix} \rightarrow G_u(4) = \begin{cases} -1.5615 \\ \\ 2.5615 \end{cases}$$

F:3.2-1

ELECTRONIC STRUCTURE OF C_{60}
Hückel 3D-MO

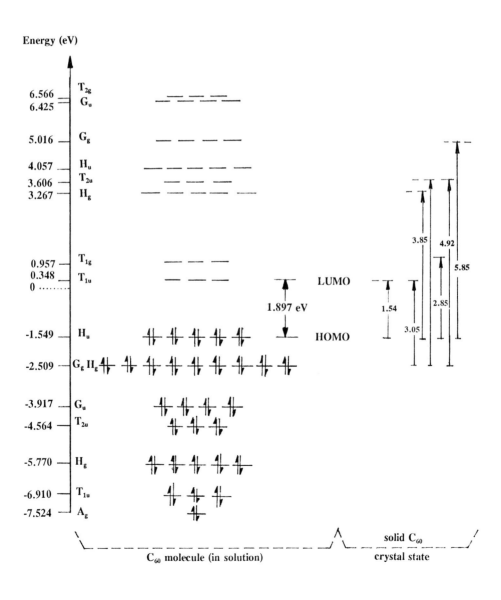

$$H_u \begin{vmatrix} x & 1 \\ 1 & x\text{-}1 \end{vmatrix} \rightarrow H_u(5) = \begin{cases} \text{-}0.6180 \\ 1.6180 \end{cases}$$

Because this calculation considers the hypothetical planar case (resonance integral of β), the non-planar case should be corrected by a factor of 0.877 as the ratio of the π bandwidth for C_{60} (3-D configuration) and graphite (2-D configuration). In real energy (unit eV) T_{1u}, as a lowest unoccupied molecular orbital (LUMO), has a value of 0.348, while H_u, as a highest occupied molecular orbital (HOMO), has a value of -1.549. This means that the energy gap $E_g = 1.897$ eV in **solution** (single C_{60}), while in **crystal state**, the E_g of C_{60} should be about 10-15% smaller.

When the Hückel calculation is applied to the π orbitals of hydrocarbons, β represents the energy of interaction among the orbitals and adjacent atoms. β is taken as the primary unit of energy because the energy of an electron in a carbon $p\pi$ orbital before interaction with other orbitals, α, is usually taken as the energy baseline. This means that β is inherently a negative quantity and all occupied molecular orbitals have negative values, while unoccupied have positive values.

The lowest excitations $H_u \rightarrow T_{1u}$ give a value of 1.897 eV, which could be split up by the influence of electron-electron interactions. This information is important for imaging by scanning tunneling microscopy (STM), because the T_{1u} unoccupied molecular orbital will interact with the substrate. The next unoccupied molecular orbital (T_{1g}) is interesting as the first optical transition $H_u \rightarrow T_{1g}$ with energy value of 2.5 eV. There are five more possible optical excitations: $H_g \rightarrow T_{1u}$, $H_u \rightarrow H_g$, $G_g \rightarrow T_{2u}$, $H_g \rightarrow T_{2u}$ and $H_u \rightarrow G_g$.

Using molecular-orbital calculations and symmetry of the icosahedron, there exists a one-to-one correspondence through the irreducible representations (A_g, T_{1g}, T_{2g} . . .) and 120 symmetry operations.

The charge distribution of the C_{60} molecule based on the HyperChem program's calculations shows that the position of carbon atoms is inside the electron cloud (Fig. 3.2-3). The diameter of C_{60} is about 1 nm, while the diameter of the position of the carbon atoms is about 0.71 nm. A picture of hybridized molecular orbitals of the C_{60} molecule is presented, too. Hexagons and pentagons are recognizable. Calculations have shown that hexagons

ELECTRONIC STRUCTURE OF C_{60}
HOMO - LUMO

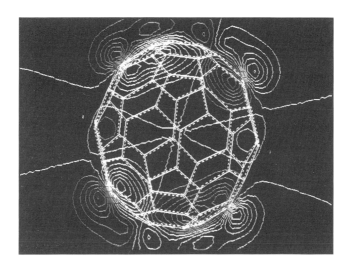

The highest occupied molecular orbital (HOMO) of the C_{60} molecule.

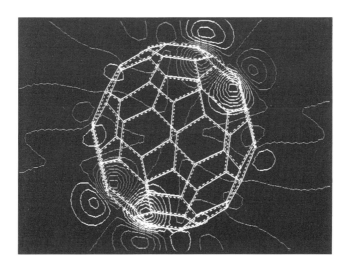

The lowest unoccupied molecular orbitral (LUMO) of the C_{60} molecule.

F:3.2-3

C_{60}
Third form of pure carbon

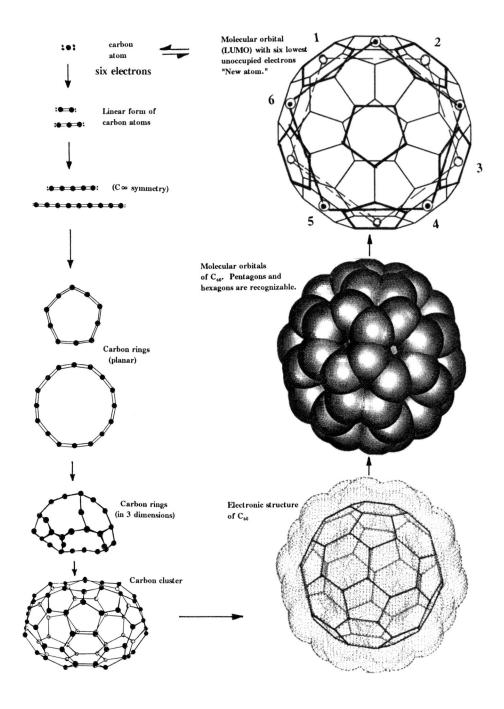

carbon atom

six electrons

Molecular orbital (LUMO) with six lowest unoccupied electrons "New atom."

Linear form of carbon atoms

$(C \infty \text{ symmetry})$

Carbon rings (planar)

Molecular orbitals of C_{60}. Pentagons and hexagons are recognizable.

Carbon rings (in 3 dimensions)

Electronic structure of C_{60}

Carbon cluster

possess "energy holes" (two visible in picture), while pentagons are closed from energy surface point of view.

The electron density of C_{60} in solution and the solid state is presented in Figure 3.2-4 (Satpathy, 1986; Saito and Oshiyama, 1991). There is a similarity of electron density in both cases. High-resolution transmission electron microscopy (HRTEM) has shown that there is a hole of electron density of C_{60} in the solid state. Theory has also predicted that electron density in the center of C_{60} in solution is very small (Satpathy, 1986), but the electron cloud is closed. The total number of electrons and electron density as a function of distance from the center of the C_{60} molecule (in solution) is presented in the same picture (Satpathy, 1986). Valence-electron density on a plane passing through the center of the cluster and two 6-6 bonds is presented by contours. The electronic structure of the C_{60} molecule is based on chemical bonding consists of sp^2 hybrid orbitals, interactions between δ and sp^2 orbitals, and $sp^2\delta$ and partial π-electron bonds between nearest-neighbor carbon atoms. For solid state of C_{60}, the contour map shows the valence-electron density with four single bonds and two double bonds with a "hole" inside of the C_{60} molecule (HRTEM picture also shows a "hole" in the C_{60} molecule).

Ring currents of the C_{60} molecule are presented in Figure 3.2-5. Calculation is based on the London approximation of the calculation of current in carbon rings (Pasquarello et al, 1992) and given as Hamiltonian matrix:

$$I_{ij} = [\sum_n (c_i^n)^* C_j^n] \exp \{ie/2hc[A(R_i)-A(R_j)]\}i_{ij} \qquad (3.1.6)$$

where: I_{ij} is current from site R_i to the nearest-neighbor site R_j, C_i^n is the eigenvectors of Hamiltonian matrix, $A(r)$ is the vector potential, and i_{ij} is the vector directed along the bond and of constant module, h is Planck's constant divided by 2π, c is the speed of light. Ring current for each bond is presented for pentagons and hexagons as the sum of $I_{ij} + I_{ji}$. Intensity of ring currents is higher in pentagons than hexagons. As we know, ring currents generate magnetic fields and calculated ring-current chemical shifts (Pasquarello et al, 1992) are presented in Figure 3.2-5 for pentagons and hexagons. Calculations show that ring currents, which are predicted by NMR experiments, are very sensitive to electronic structure. Elser and Haddon (1987a and 1987b) calculated the magnetic behavior of the C_{60} molecule and ring current magnetic susceptibility based on London theory of the magnetic response of aromatic hydrocarbons. They concluded "that the π-electron ring-current susceptibility

ELECTRON DENSITY OF C_{60}
                         (in solution and solid state)

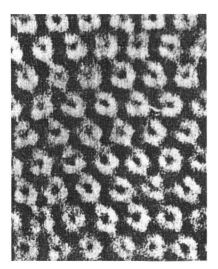

High-resolution transmission electron
microscopy (HRTEM) of solid C_{60}.

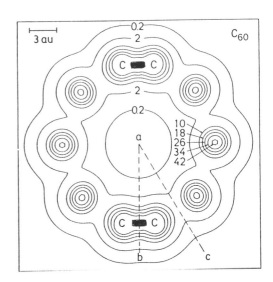

Electron density of the C_{60} molecule
(in solution)

The valence-electron density of
the solid C_{60} (crystal on the [010]
plane)

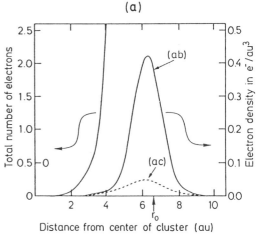

Total number of electrons and
electron density as a function
of distance from center of cluster
for C_{60} molecule.

F:3.2-5

RING CURRENTS OF C$_{60}$

Ring current in pentagons

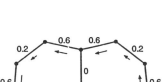

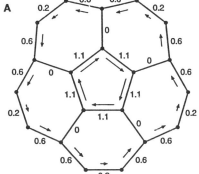

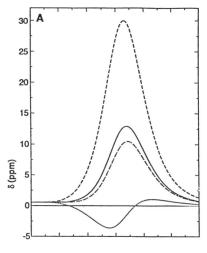

Ring current in hexagons

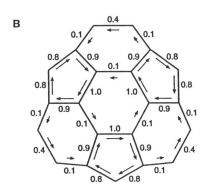

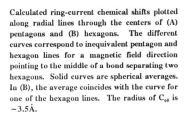

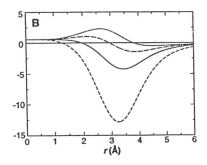

Calculated ring-current chemical shifts plotted along radial lines through the centers of (A) pentagons and (B) hexagons. The different curves correspond to inequivalent pentagon and hexagon lines for a magnetic field direction pointing to the middle of a bond separating two hexagons. Solid curves are spherical averages. In (B), the average coincides with the curve for one of the hexagon lines. The radius of C$_{60}$ is ~3.5Å.

Electron ring current in C$_{60}$ for a magnetic field oriented perpendicular to a plane containing (A) a pentagon and (B) a hexagon, respectively. The currents in the pentagons are paramagnetic. The current strength is given with respect to that in benzene.

of C_{60} is unusually small and sensitively dependent on the relative strengths of the two inequivalent bonds of the molecule." Because the C_{60} molecule rotates *randomly*, both in solution and in the solid state, doping (inside and outside of the C_{60} molecule) and technological improvement (dimer with oriented rotation) may implement use of Fullerene C_{60} as a basic material for electronic devices.

3.3 C_{60} Vibrational Structure

Generally, the energy of a C_{60} molecule consists partly of electronic energy, vibrational energy, rotational energy and translational energy. Molecular vibrations give rise to absorption bands through the infrared (IR) and Raman (R) regions of the electromagnetic spectrum. Electromagnetic radiation is characterized by its wavelength (λ), its frequency (f), and wavenumber υ. Relationships among parameters are:

$$\upsilon = \frac{f}{c/n} = \frac{1}{\lambda} \ [cm^{-1}] \qquad (3.3.1)$$

where: c is the velocity of light in a vacuum (3×10^{10} *cm/s*) and (c/n) is the velocity of light in a medium whose refractive index is n, in which the wavenumber is measured. The wavenumber is sometimes referred to as the "frequency in cm^{-1}" or just "frequency," which creates confusion between frequency "*in Hz*" and "frequency *in cm^{-1}*." The frequency in *Hz* in the IR part of the spectrum is an inconveniently large number and the wavenumber (cm^{-1}) is more commonly used. In Raman spectroscopy only the wavenumber is used.

In the study of molecular vibrations of the C_{60} molecule, we have started with a classical model of the molecule where the carbon atoms are represented by mathematical points with mass. It is well known that *3N-6* internal degrees of freedom of motion of a nonlinear molecule correspond to the same number (*3N-6*) of independent normal modes of vibration. For the C_{60} molecule, there are $3 \cdot 60 - 6 = 174$ active modes. To understand vibrational spectra of the molecule, it is very important to have a good familiarity with the molecule's symmetry, because in each normal mode of vibration all the atoms in the molecule vibrate with the same frequency and all the atoms pass through their positions simultaneously. Based on the symmetry of the C_{60} molecule (Equation 2.3.1 and F:2-21), we know which *Ih* symmetry group (irreducible representations--A_g, T_{1g} ...) should be used. For vibration in the IR domain,

F:3.3-1

VIBRATIONAL SPECTRA OF C_{60}

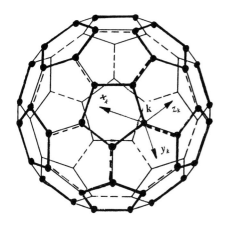

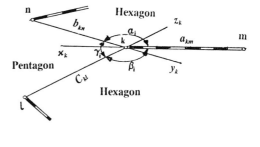

$$3N - 6 = 3 \cdot 60 - 6 = 174$$

Number of vibration modes of C_{60}

I_h	# Eigenvalue	irrep dimensionality	Total number of modes
A_g	2	1	2
T_{1g}	3	3	9
T_{2g}	4	3	12
G_g	6	4	24
H_g	8	5	40
A_u	1	1	1
T_{1u}	4	3	12
T_{2u}	5	3	15
G_u	6	4	24
H_u	7	5	35

A_g and H_g - Raman (10)
T_{1u} - Infrared (4)

Symmetry labelled eigenfrequencies of C_{60}

Even parity		Odd parity	
I_h group label	„ Frequency " (1/cm)	I_h group label	„ Frequency " (1/cm)
A_g	1830	A_u	1243
	510		
		T_{1u}	1868
T_{1g}	1662		1462
	1045		618
	313		478
T_{3g}	1900	T_{3u}	1954
	951		1543
	724		1122
	615		526
			358
G_g	2006		
	1813	G_u	2004
	1327		1845
	657		1086
	593		876
	433		663
			360
H_g	2085		
	1910	H_u	2086
	1575		1797
	1292		1464
	828		849
	526		569
	413		470
	274		405

we should look for T_{1u} irreducible representation. There are 12 number of modes for this representation, but its irrep dimensionality is 3, so that there are only 4 eigenvalues for the C_{60} molecule in the IR domain. This fact was crucial to Huffman/Krätschmer when they produced the C_{60} molecule in the solid state (Krätschmer, 1990). Their x-ray diffraction pattern of C_{60} and electron diffraction (dots in the upper left) are presented in Figure 3.3-2. The C_{60} molecule crystallizes by the T_h symmetry group as "fcc" lattice (see Figure 2-16).

One of the first results of vibrational frequencies for the C_{60} molecule came from *ab inito* calculation using the STO-36 approach (Disch and Schulman, 1986).   In the same year, Newton and Stanton used MNDO (Modified Neglect of Diatomic Overlap), and they reported that the range of wave number of molecular vibrations of the C_{60} molecule was from 186 to 1217 cm^{-1}. A year later, Wu, Jelski and George (1987) presented vibrational analysis of the C_{60} molecule based on a 180x180 matrix.   Experimental (neutron scattering and high-resolution electron-energy-loss spectroscopy) and theoretical (semi-empirical quantum-chemical calculation) are presented by Negri, Orlandi and Zerbetto (1992).

For vibration in the Raman domain, important irreducible representations are A_g and H_g. They give a total number of 42 modes, but their irrep dimensionalities are 1 and 5, respectively, with only 10 eigenvalues. The number of vibrational modes of the C_{60} molecule with their values is summarized and presented in Figure 3.3-1.

During the vibration, the bond lengths a_{km}, b_{kn} and c_{kl} change the angles α, β, and γ due to the motion of the atoms and give (T$_i$) as the kinetic energy of each atom (Wu et al, 1987).

The variation of the bond lengths ($\delta\, a_{km}$, $\delta\, b_{kn}$ and $\delta\, c_{kl}$) are the first order (Figure 3.3-1):

F:3.3-2

SOLID STATE OF C_{60}

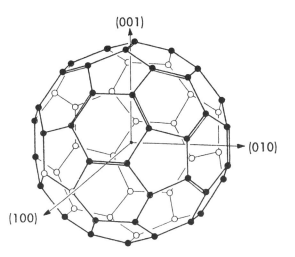

C_{60} crystal axes of the *fcc* lattice.

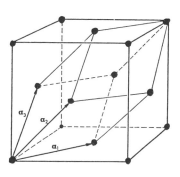

T_h symmetry group as *fcc* lattice

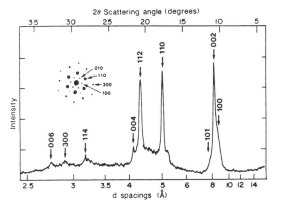

X-ray diffraction pattern of solid
C_{60} and electron diffraction (upper left).

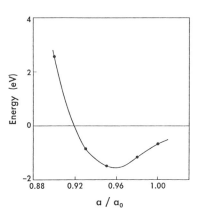

Total energy per C_{60} in the *fcc* C_{60}
solid as a function of the lattice
constant a (a_0 is an experimental lattice
constant)

$$\delta a_{km} = (x_k - \frac{1}{2}z_k) + (x_m - \frac{1}{2}y_m - \frac{1}{2}z_m)$$

$$\delta c_{kl} = [(z_k - \frac{1}{2}x_k - \cos(\frac{2}{5})\pi y_k) + (y_l - \frac{1}{2}x_l - \cos(\frac{2}{5})\pi z_l)]$$

$$\delta b_{kn} = [(y_k - \frac{1}{2}x_k - \cos(\frac{2}{5})\pi z_k) - (\frac{1}{2}x_n - \cos(\frac{2}{5})\pi y_n)]$$

The angle between a double bond and its adjacent pentagonal face is:

$$\psi = \cos^{-1}(\frac{1}{2}\cos(\frac{3}{5})\pi).$$

The angle between two adjacent hexagonal faces is:

$$\phi = \cos^{-1}[\frac{8}{3}(\sin\frac{3}{10}\pi)^2 - 1].$$

The angle between adjacent hexagonal and pentagonal faces is:

$$\theta = \frac{1}{2}(\pi - \phi) - \psi$$

Variations (δ) of the angles between bonds α, β and γ are:

$$\delta\alpha_i = (\frac{1}{L})[-\sin(\frac{1}{3})\pi \cos \phi y_m + \sin(\frac{1}{3})\pi x_n -$$

$$\sin(\frac{2}{5})\pi \cos \theta y_n - \sin(\frac{1}{3})\pi y_k + \sin(\frac{1}{3})\pi \cos \phi z_k -$$

$$\sin(\frac{1}{3})\pi x_k + \sin(\frac{2}{5})\pi \cos \theta z_k$$

$$\delta\beta_i = (\frac{1}{L})[-\sin(\frac{1}{3})\pi \cos \phi z_m + \sin(\frac{1}{3})\pi y_m +$$

$$\sin(\frac{1}{3})\pi x_L - \sin(\frac{2}{5})\pi \cos \theta z_l - \sin(\frac{1}{3})\pi z_l +$$

$$\sin(\frac{1}{3})\pi \cos \phi y_m - \sin(\frac{1}{3})\pi x_m +$$

$$\sin(\frac{2}{5})\pi \cos \theta y_m]$$

$$\delta\gamma_i = (\frac{1}{L})[\sin(\frac{2}{5})\pi y_n - \sin(\frac{1}{3})\pi \cos \theta x_n +$$

$$\sin(\frac{2}{5})\pi z_l - \sin(\frac{1}{3})\pi \cos \theta x_l + \sin(\frac{1}{3})\pi \cos \theta x_k -$$

$$\sin(\frac{2}{5})\pi y_k + \sin(\frac{1}{3})\pi \cos \theta x_k - \sin(\frac{2}{5})\pi z_k]$$

The kinetic energy of each atom is:

$$T_i = (\frac{1}{2})m\{[\frac{\partial x_i}{\partial t} - (\frac{1}{2})\frac{\partial y_i}{\partial t} - (\frac{1}{2})\frac{\partial z_i}{\partial t}]^2 + [\frac{\partial y_i}{\partial t} \cdot \cos(\frac{1}{5})\pi -$$

$$- \frac{\partial z_i}{\partial t} \cdot \cos(\frac{1}{5})\pi]^2 + [\frac{\partial y_i}{\partial t} \cdot \cos(\frac{3}{10})\pi \cdot \sin \psi +$$

$$+ \frac{\partial z_i}{\partial t} \cos(\frac{3}{10})\pi \sin \psi]^2\}.$$

The total number of eigenvalues of the C_{60} molecule is 46. There are 5-10% differences in the values of both experimental and theoretical data. These differences are a consequence of different models and different experimental conditions. One of the possible values of 46 eigenvalues of the C_{60} molecule is given in Figure 3.3-1. The following literature citations are most relevant: Krätschmer et al, 1990; Berthune et al, 1990; Berthune et al, 1991; Frum et al, 1991; Sinha et al, 1991; Tolbert et al, 1992, Jiang et al, 1992; Klug et al, 1992; Suresh et al, 1992; and others.

CHAPTER 4

STM IMAGES AND
VIBRATIONAL-ROTATIONAL SPECTRA OF C_{60}

4.1 STM Images of C$_{60}$ Molecules: An Approach to the Nanoscale

4.1.1 STM: Basic Principles and A Short History

Functional communication among biomolecules and technological devices including Fullerenes will require dramatic advances in our abilities to fashion logic devices from matter. The attainable limit appears to be components structured at the atomic level, on the order of 0.1-0.3 nanometers. Nanoscale fabrication of information devices capable of biological interfacing would also enable construction of valuable nanoscale robots, sensors, and machines (Schneiker, et al, 1988; Hameroff, 1987). Fullerene technology may be an important facet of a wide range of applications: "nanotechnology".

The American Physical Society held its 1959 annual meeting at the California Institute of Technology. Scheduled to speak was a man who would win the 1965 Nobel physics prize for his historic work in quantum electrodynamics. Still later he would serve on a Presidential Commission and find the fault in the Challenger disaster in a disturbingly brief period of time. In 1959 however, he spoke of other work that may be even more important. In his talk entitled *There's Plenty of Room at the Bottom*, physicist Richard Feynman (1961) proposed a simple and straightforward strategy for constructing useful structures ranging in size down to the atomic scale! He suggested using machine tools to make many more sets of much smaller machine tools, which would in turn make many times that number of other, even smaller machine tools, and so on. At the lowest level, he noted the possibility of mechanically assembling molecules, *an atom at a time*.

Feynman proposed construction of nanomachines, nanotools, tiny computers, molecular scale robots, new materials and other exotic products which would have far reaching applications and benefits. Considering the relevance of nanotechnology to living molecules, Feynman (1961) noted, "A biological system can be exceedingly small .... Consider the possibility that we too can make a thing very small, which does what we want-that we can manufacture an object that maneuvers at that level!!"

Machines able to directly manipulate matter on the submicron to nanometer size scale are referred to as "Feynman Machines" (FMs). The only existing FMs currently known are biological, however computer controlled or teleoperated FMs may in the future implement a broad range of nanotech applications.

Following Feynman's lead, other scientists delved into nanotechnology.

Von Hippel (1962) predicted dramatic material science possibilities if new advances in "molecular designing" and "molecular engineering" of materials could be achieved. Noting the eventual possibility for repairing human tissue (molecule by molecule if necessary) for life extension, Ettinger (1964) suggested repair machinery for modification and interaction with existing organisms; later he proposed development of nanorobotics. Ettinger envisioned nanoscale scavenger and guardian organisms designed to emulate and surpass the actions of white blood cells which might hunt down and clean out hostile or damaging invaders (Ettinger, 1972). Such nanorobots might be useful for fighting AIDS or cancer, excavating blocked blood vessels, or straightening neurofibrillary tangles in senile (Alzheimer) neurons.

Shoulders (1965) reported the actual operation of micromanipulators able to position tiny items with 10 nm accuracy under direct observation by field ion microscopy. Ellis (1962) also developed similar (but much larger) micromanipulators and proposed the construction of "microteleoperators": remote controlled nanodevices! Drexler (1981, 1986) has described some advantages and hypothetical dangers of nanotechnology. Capabilities for atom-by-atom assembly and nanoengineering could lead to new materials and pathways (Feynman, 1961). One such material is "diamond-like carbon" films which are "transparent, insulating, chemically inert, have a high dielectric strength, good adhesion and are relatively hard" (Aisenberg, 1984).

Potential benefits from nanotechnology attainable in perhaps a decade or two might include (Schneiker, et al, 1988): "... vastly faster, much more powerful and numerous computers with extremely large capacity memories, ultrastrong composite materials, greatly improved scientific instrumentation, microscopic mobile robots, and automated flexible manufacturing systems, replicating systems, and achieving the practical miniaturization limits and maximum performance in virtually every area of technology."

Despite these lofty hopes and predictions, nanotechnology and molecular computing have remained mere dreams. Obstacles to their implementation center on the absence of available Feynman machines. A feasible solution was advanced by Conrad Schneiker, who predicted, building on Feynaman's ideas, a bridge to the nanoscale. In 1986, Schneiker suggested that atomic level manipulative capabilities embodied in a 1981 invention, the scanning tunneling microscope (STM), can implement nanoscale Feynman machines (Hansma and Tersoff, 1987).

SCANNING TUNNELING MICROSCOPES (STMs)

Mode Number	I	II	III	IV
Quantity held constant	i, v	h, v	h, i	h, i, v
Quantity measured	h	i	v	i/v

Table 4.1.1-1: STM operating modes; i, v, and h are tunneling current, voltage and height (tip to sample surface distance), respectively.

The 1986 Nobel Prize for Physics was split between the 50 year old cornucopia of microknowledge, the electron microscope, and a brand new, unfulfilled technology, scanning tunneling microscopy (STM). The STM half of the Nobel Prize was awarded to Gerd Binnig and Heinrich Rohrer of the IBM - Zurich Research Laboratories where STM was invented in 1981 (Binnig, Rohrer, Gerber and Weibel, 1982). The principle used in STM is extremely simple. Piezoceramic materials expand or shrink extremely small and precise distances (i.e. angstroms, or tenth nanometers) in response to applied voltage. In STM, piezoceramic holders scan an ultrasharp conducting tip (such as tungsten) over a conducting or semiconducting surface. When the STM tip is within a few angstroms of a surface, a small voltage applied between the two gives rise to a tunneling current of electrons. The tunneling current depends exponentially on the tip-to-substrate separation (about an order of magnitude per angstrom or tenth nanometer). Depending on the substrate, typical tunneling currents and voltages are on the order of nano-amperes and millivolts, respectively. A servo system uses a feedback control that keeps the tip to substrate separation constant by modulating the voltage across a piezoelectric positioning system. As the tip is scanned across the surface, variations in this voltage, when plotted, correspond to surface topography. Detailed surface topography maps of various materials have been obtained which demonstrate individual atoms like cobblestones. By varying and measuring different combination of tunneling current, voltage, and distance, different types of information may be obtained about the surface being probed.

The basic modes of operation of an STM, described by Hansma and Tersoff (1987), are summarized in Table 4.1.1-1. Here i, v, and h are the tunneling current, the voltage across the gap, and the gap size respectively.

Mode I is used to measure the topography of the surface of a metal or semiconductor and is the slowest mode since the electro-mechanical servo system must follow the shape of the surface during the scanning operation. The scanning speed in Mode I is determined by the response of the servo system. Modes II and III are faster since the tip maintains only a constant average height above the surface. The scanning speed in these modes is determined by the response of the preamplifier only. Mode IV measures the joint density of states which, for a small tip, is a map of the local distribution of electron states of the substrate. For this mode, one varies the tip-to-substrate bias voltage with a small alternating current signal and monitors i/v. Using this mode, Smith and Quate (1986) have done spatially resolved tunneling spectroscopy and observed charge density waves. STM can thus be used to identify the elements comprising specific atoms and to monitor molecular dynamics.

Adaptability and versatility have been shown by STMs operated in air, water, ionic solution, oil and high vacuum (Drake, Sonnenfeld, Schneir, Hansma, Solugh and Coleman, 1986; Miranda, Garcia, Baro, Garcia, Pena and Rohrer, 1985). Their scanning speed may be pushed into the real-time imaging domain (Bryant, Smith and Quate, 1986). One technique for machining STM tips is based on a simple ion milling process that can generate ultrasharp tips with single atom points; a similar technique can generate ultrasharp knife edges and other nanotool shapes as well (Dietrich, 1984). Dieter Pohl (1987) of IBM-Zurich considers STM one example of a group of "stylus" microscopic technologies. Tips are being developed as thermocouples of two different metals sensitive to extremely low changes of heat, and hollow pipettes able to administer or detect individual molecules in solution. Pohl refers to these STM capabilities as molecular "tasting and smelling." Another STM related function can "touch and feel." In "atomic force microscopy," a nanoscale lever positioned by STM piezo-technology is deflected by contact and/or movement of atoms, molecules or their surrounding ions. Van der Waals force induced lever deflection is monitored by an STM tip or by interferometry. Binnig, Quate, and Gerber (1986) propose to measure forces as small as 10^{-18} newtons using atomic force microscopy. Mechanical shape and structural dynamics can be probed without tunneling through the sample material using this mode, which may be particularly important in the study of biomolecules. Other STM tip adaptations (ionic sensors, GigaHertz resonators, etc.) have generalized the technology so it is known as either scanning probe microscopy (SPM) or scanning (variable) tip microscopy (STM or SXM).

Another STM spinoff may lead to "nanovision." Resolution of "near-field microscopy" depends on the diameter of the aperture through which the reflected light passes. Binnig (1985) described how to use an STM to make 20 nm holes in opaque metal films which, when used in scanning light microscopes, become capable of resolving features much smaller than the wavelength of the light. In general, a cone of light cannot be focused to a spot which is significantly smaller than the light's wavelength. However, light passed through an aperture that is many times smaller in diameter than its wavelength results in a narrow beam of light that may be scanned mechanically close to a sample being examined. STM technology is capable of creating nanoapertures, and also the scanning mechanism required for imaging an area. Pohl, Denk and Lanz (1984) described their "optical stethoscope" in which "details of 25-nm size can be recognized using 488-nm radiation." A significant advantage of the optical stethoscope concept is "that it allows samples to be *nondestructively* investigated in their *native* environments" (Lewis, Isaacson, Harootunian and Muray, 1984). Betzig and colleagues (1986) conclude that this technology will be "able to follow the temporal evolution of macromolecular assemblies in living cells...to provide both kinetic information and high spatial resolution." Scanning nanoapertures could also use ultraviolet, X-ray, gamma ray, synchotron, or particle beams to "snoop" in the nanoscale if suitable materials and configurations are found.

Rudimentary nanomanipulation has already been described. Ringger and colleagues (1985) have described STM nanolithography: atomic scale surface etching. Becker, Golovchenko and Swartzentruber (1987) of Bell Labs have reported a single atom modification of the surface of a nearly perfect germanium crystal by the tungsten tip of an STM. They conclude that manipulation of atomic sized structures on surfaces will lead to "new frontiers in high density memories, new devices whose electrical properties are dominated by quantum-size effects, and modification of single, self replicating molecules." Foster and Frommer (1989) used an STM tip to move and then cleave a single pthallocyanine molecule on an atomic surface. Eigler (1992) used an STM tip to construct a single atom (xenon) "switch". In its first decade, STM has imaged single atoms, probed atomic forces, manipulated atoms, and won a Nobel prize. STM has become an "interface between our microscopic world and the world of nano-objects, possibly in their natural environment" (Rohrer, 1992). Eventually STMs and their spinoffs may enable researchers to alter and communicate with genes, viruses, proteins, cytoskeletal assemblies, Fullerene devices, and perhaps biomolecular awareness.

Fullerenes, themselves predicted to have piezoelectric properties, may be implemented as nanoscale STM devices.

STM research and development began at the University of Arizona in 1983, two years after STM invention. Conrad Schneiker, seeking to implement ideas, approached Stuart Hameroff about building an STM to study and interface with cytoskeletal microtubules.   Schneiker, working with fellow Optical Sciences graduate student Mark Voelker in the laboratory of Eugtace Dereniak, constructed a working STM and obtained early papers on STM imaging of biomolecules (Voelker et al, 1988). Later Dror Sarid and others accrued commercial devices in an STM/AFM facility, and Voelker developed a multi-tip STM for his Ph.D. thesis.

4.1.2 Imaging Fullerene C_{60} by STM

Images with internal and atomic resolution of Fullerene C_{60} have been made by Koruga et al (1993).   Mixed molecular weight Fullerenes were synthesized using the Huffman/Krätschmer procedure (Krätschmer et al, 1990), and C_{60} was purified using a chromatographic procedure (MER Corporation, Tucson, Arizona, USA). A three-layered substrate was designed and prepared; the upper layer, a gold substrate (Au), was sputter cleaned and imaged by a NanoScope II STM (Digital Instruments Inc., Santa Barbara, California, USA). In control studies of the cleaned gold layer by STM in air, no structures or irregularities were identified. Before being applied to the gold layer, the dry purified Fullerene C_{60} samples were redissolved in toluene. The STM tip was mechanically cut from Pt/Ir wire and the experiment was done on three samples at room temperature. The NanoScope II parameters for tunnelling current and bias voltages were 0.1-5 nA and 0.02-1 V (Z=20.0 Å/V, XY=18.2 Å/V) with effective "gap resistance" of about 10^8 ohm. The acquisition time for the image was 20 sec (400/scan). Under certain conditions (three layers, electro-magnetic and mechanical properties of C_{60}, bias voltage of STM), C_{60} was fixed on the substrate (no movement, no rotation) and its internal structure observed. Optimal parameters for imaging Fullerene C_{60} with internal structure resolution was 1.0 nA (tunnelling current) and 20.1 mV (bias voltage). Previous reports of STM imaging of Fullerenes C_{60} and C_{70} (Wilson et al, 1990; Li et al, 1991; Zhang et al, 1992; Wragg et al, 1990; Lamb et al, 1992) failed to show clear internal structure and atomic resolution.

Fullerenes in solution rotate (rotational diffusion constant $D = 1.8 \times 10^{10}$

F:4.1.2-1

C_{60} MOVEMENTS ON SUBSTRATE AND STM IMAGE

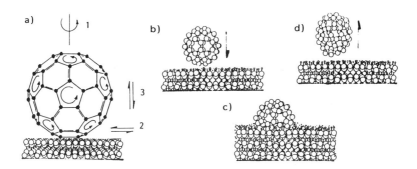

The C_{60} molecule has three main movements in solution on substrate: rotation; sliding; and bouncing. STM image of C_{60} molecule with internal structure, including atomic resolution, is possible only in position "c" (no-rotation, -sliding, -bouncing; same kind of "frozen state").

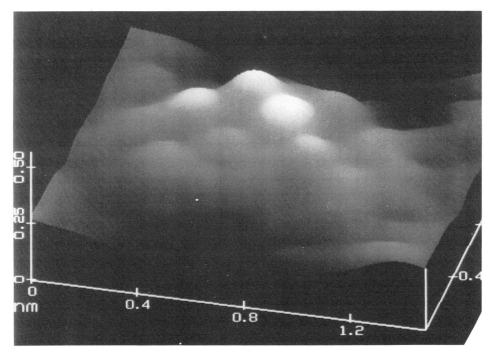

STM image of C_{60} molecule in position "c" under certain conditions (electro-magnetic and mechanical properties of C_{60}, substrate, and bias voltage of STM).

CLUSTERS OF C_{60}/C_{70}
STM Images

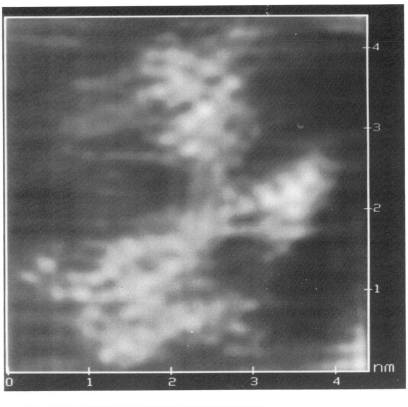

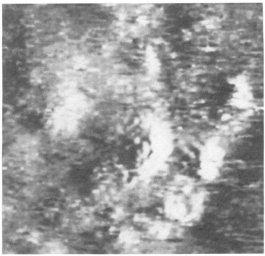

Scanning Tunnelling Microscope (STM) images of mixed Fullerenes C_{60}/C_{70}. There are interactions between Fullerenes, but the nature of the interactions is yet unknown. The upper layer of the substrate is gold. The experiment was done at room temperature in air.

per second) and move around and bounce on the substrate (speed near 3×10^5 cm/s) as shown in Figures 4.1.2-1a, b, c, and d (Johnson et al, 1992; Beck et al, 1991). All pentagonal rings rotate in the same direction (Stanton and Newton, 1988), while the "ring current" (delocalized π electron) during Fullerene rotation generates a magnetic field. When we used a two-layer substrate (gold-silicon), the Fullerene movement and bouncing on the substrate was muffled [through hybridization of the highest occupied molecular orbital (HOMO) and lowest unoccupied molecular orbital (LUMO) of the Fullerene and HOMO of the golden substrate] and we observed Fullerene structure as in Figure 4.1.2-2. Under these conditions, it is not possible to observe the internal structure of the Fullerenes (the Fullerene rotates and bounces slightly). Based on magnetic susceptibility data (Haddon and Elser, 1990) of C_{60} and the STM current, we predicted the effects of magnetic induction of the substrate on C_{60}. With a three-layer substrate, C_{60} movement was eliminated (through high level hybridization of molecular orbitals of Fullerene and the substrate, resulting in electromagnetic effects of the alloy layer on Fullerene and its mechanical deformation) such that C_{60} was "stuck on" the substrate (Figure 4.1.2-1c).

Using a three-layer substrate (silicon-alloy-gold) Fullerene was fixed on the substrate (Figure 4.1.2-1c and its STM image) and the internal structure of an individual Fullerene revealed as pentagons and hexagons (Figure 4.1.2-3). Comparing Figures, we can see that pentagons appear white and hexagons black. The STM constant current mode views primarily the high-lying π electrons on the molecular energy surface of C_{60}. This suggests that the molecular energy of C_{60} is concentrated in the (white) pentagons. Surface energy images and structural images differ slightly. Because two carbon atoms are missing (7 and 11), the energy surface is broken (Y). This suggests that Fullerene C_{60} presented in Figure 4.1.2-3 has a defect and is missing two carbon atoms, C_7 and C_{11} connected by a double bond, resulting in C_{58}. The minimum energy necessary for the dissociation process ($C_{60} \rightarrow C_{58} + C_2$) is 4.6 eV (Radi et al, 1990). The center pentagon (C_1-C_5) in the STM picture is distorted compared to the stick model of Fullerene C_{60}. This could be due to a surface defect Y and an energy dislocation. The energy value for C_{60} is 0.4383 eV/atom (Adams et al, 1992), while for C_{58} (C_{60} with defect) it is much higher: 0.7168 eV/atom. This energy value is based on quantum molecular dynamics calculations (Sankey and Niklewski, 1989; Adams et al, 1991) as approximate energies per atom (relative to single-plane graphite) of Fullerene

F:4.1.2-3   FULLERENE C_{60}/C_{58}
                             STM image

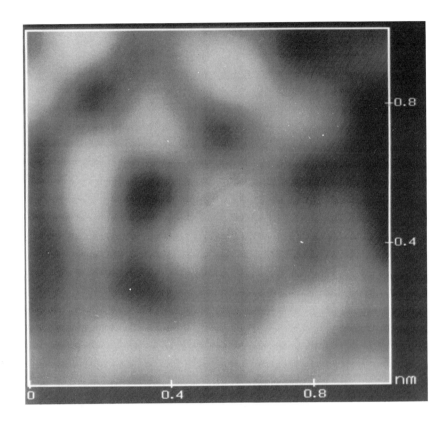

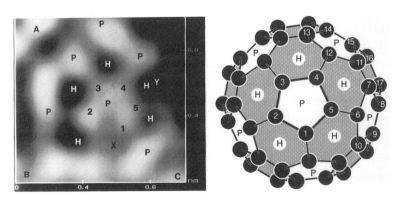

First STM image of C_{60}/C_{70} with internal structure: pentagons (white - high level surface energy concentration) and hexagons (black - "energy hole"). When two atoms, C_7 and C_{11} are missing, there is a defect Y and the C_{60} molecule becomes C_{58} (or C_{60} with a big hole--open structure).

F:4.1.2-4   ATOMIC RESOLUTION OF C_{60}
                              STM image

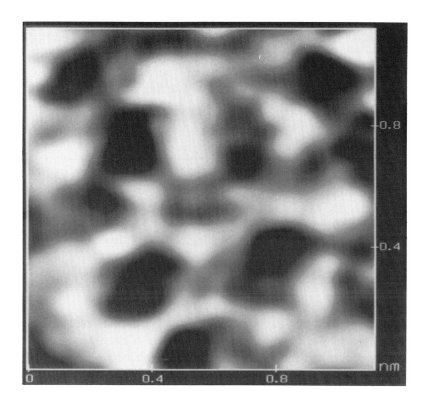

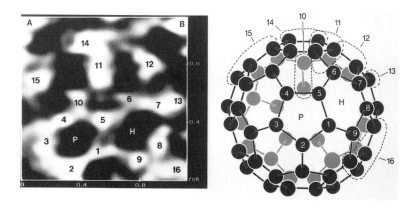

An atomic resolution image of C_{60} by STM. Rings of five and six carbon atoms (the high level energy concentration of molecular orbitals) are visible. Other parts of the image are less clear because it was high-level hybridization of molecular orbitals of the C_{60} molecule and substrate. Comparative stick model of C_{60} is presented and regions are marked to show the same details.

molecules, and the average π-angle of the molecule assuming a perfect sphere. Because the Fullerenes were fixed on the substrate ("frozen state"), C_{58} has the basic structure of C_{60} (C_{58} is C_{60} with a hole).   We believe that the STM tip may have severed the bonds during imaging, because if C_{58} occurred during synthesis, it seems more likely to appear "closed," rather than as C_{60} with a hole.

Figure 4.1.2-4 shows an image of C_{60} with diameter about 1 nm which has been low-pass filtered (to reduce high-frequency noise). The STM image shows clearly one pentagon and one hexagon carbon ring of C_{60}. The other parts of the C_{60} image are fuzzy. This may be due to interaction of the C_{60} with the three-layer substrate ("frozen state"), the position of C_{60} on the substrate and/or its relationship with the STM tip during imaging. Figure 4.1.2-4 shows a hexagonal "energy hole" (under C_4-C_5) through which electrons can tunnel into the C_{60} molecule. This could explain the view of two carbon atoms from the opposite side of C_{60} (see position 10). The hexagonal "energy hole" may be used as a "door" to introduce small atoms like hydrogen or ions for doping of C_{60} by electrochemical methods or STM techniques.

Electron affinity of C_{60} based on the photoelectron spectrum is given by the threshold energy for the appearance of photoelectrons (Curl and Smalley, 1988). Based on these data, a rather large difference between HOMO and LUMO occurs between 1.7 - 1.9 eV. This result indicates that bias voltages for STM imaging of the isolated C_{60} should be about 1 V. However, our experimental result indicated that the optimal bias voltage for STM imaging of C_{60} on the three-layer substrate with internal structure resolution was 0.0201 V. This discrepancy may be explained by the interaction between C_{60} and the substrate.   The energy separation between HOMO and LUMO has been calculated for an *isolated* Fullerene (Curl and Smalley, 1988), while the results of a much lower STM bias voltage are for *non-isolated* C_{60}. Thus hybridization of the HOMO and LUMO of the C_{60} and the HOMO of the substrate can account for electrons tunnelling *into* and *out of* the C_{60} with a lower STM bias voltage, and account for atomic resolution observation of C_{60} molecules. These experimental results demonstrate that internal and atomic resolution of C_{60} is possible at room temperature under certain conditions. Based on these findings, we foresee STM-based Fullerene information nanotechnology which, combined with electrochemical methods, may lead to self-assembling molecular computers.

STM imaging of the C_{60} molecule in position "c" (Figure 4.1.2-1) as a "frozen state" opens up the possibility of reconstructing the whole molecule.

F:4.1.2-5   **WELL-ORDERED CLUSTER OF C_{60}/C_{70}**
                          STM Images

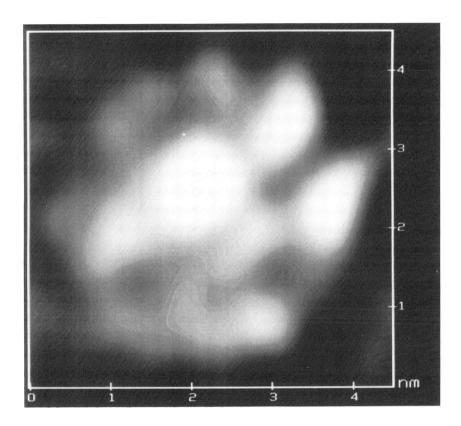

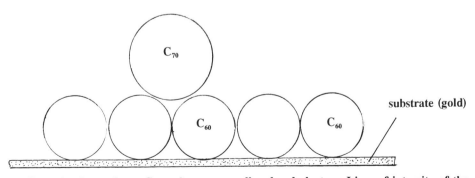

Twelve C_{60} molecules and one C_{70} make a very well ordered cluster. Lines of intensity of the electrical field are visible. Molecular forces between twelve C_{60} on the substrate are unknown. There is one C_{70} on top of the $(C_{60})_{12}$ cluster.

F:4.1.2-6

ICOSAHEDRAL SYMMETRY OF C_{60} AND STM IMAGE

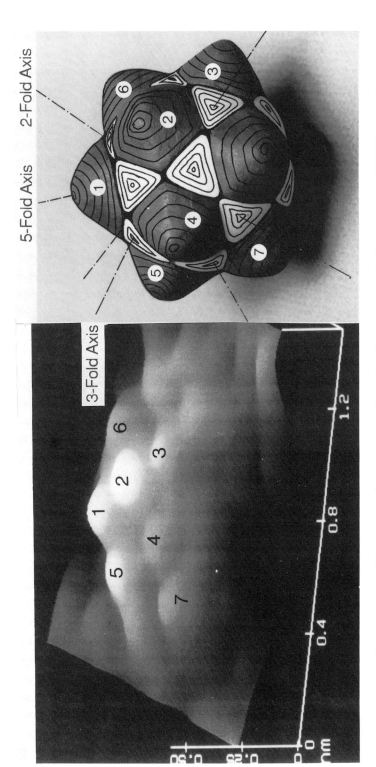

STM image of C_{60} molecule in "frozen state" shows that this molecule has icosahedral symmetry. STM images give us the first visual picture as proof that C_{60} possesses icosahedral symmetry. Because of hybridization of the HOMO and LUMO of the C_{60} with the HOMO of the substrate, only four pentagons of C_{60} are visible. Reconstruction of the whole molecule of C_{60} based on the STM image is given as a model at right.

F:4.1.2-7   **SURFACE ENERGY OF C_{60}**
                              Parameter characterization

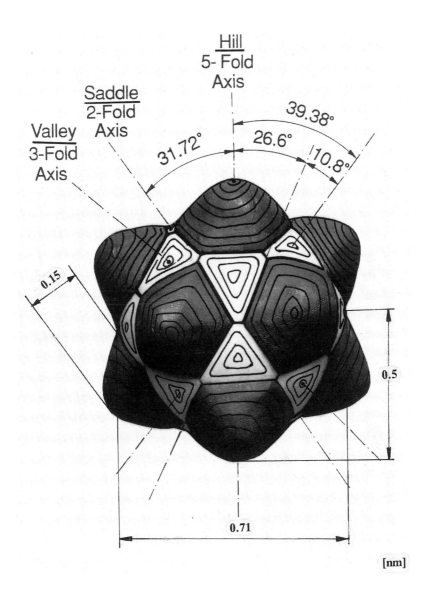

Surface energy of C_{60} is synergy output of electronic, vibration and rotation properties of this molecule. There are 12 hills, 20 valleys ("energy holes") and 30 saddles. Because the C_{60} is dynamical, these parameters are an "equilibrium state." Because of C_{60} molecule pulsing, pentagons and hexagons become bigger and smaller, and values of parameters are relative.

There are seven visible hills, which represent pentagons. Distributions of pentagons in the picture are based on icosahedron symmetry. Because C_{60} has 12 pentagons, reconstruction of the whole molecule is as given in Figure 4.1.2-6 (right picture). Based on the STM image, it is possible to characterize the C_{60} molecule (Figure 4.1.2-7). There are 12 pentagons (hills--fivefold axis), 20 hexagons (valleys--threefold axis) and 30 link points (saddles--twofold axis). Angles between symmetry axes are: 39.38°, 31.72°, 26.6° and 10.8°, as shown in Figure 4.1.2-7. The STM image gives the value of hill height of about 0.15 *nm*, diameter of 1 *nm* (based on the hills) and diameter of 0.71 *nm* (based on saddles). Atom positions are on valleys and saddle levels, meaning 0.15 nm from hilltop to center of the C_{60} molecule. These values are relative because the C_{60} molecule is dynamical and pentagons and hexagons become bigger and smaller.

The STM technique has been used for systematic study of the nucleation/growth and stability of Fullerene C_{60} films on gold as a function of coverage and temperature. This study shows relative strength of the C_{60}-Au interaction: "The C_{60} overlayer desorbs from Au(111) at a temperature which is 200° higher than the temperature required to desorb C_{60} from the second layer or to sublime it from the bulk" (Altman and Colton, 1992). The STM study indicates that two ordered structures predominate on the Au(111) surface. The first structure is a layer with a periodicity of about 38 gold spacings, while the second is a layer with a $2\sqrt{3}$ x $2\sqrt{3}$ R30° unit cell. In both cases, Fullerenes (C_{60}) were in some places brighter than others. Dynamic behavior of solid C_{60} on the substrate (rotation, sliding, and bouncing) and its catalytic chemistry with gold may explain this phenomenon.

4.2 π-Electrons in 3-D

Fullerene C_{60} is an opportunity and challenge for researchers to make a "mental jump" from planar (two-dimensional) geometrics to three-dimensional representation of π-electrons. Hadon (1988) is a pioneer in research on π-electrons in general and on the C_{60} molecular in particular. Based on π-orbital axis vector (POAV), he has developed a method for calculating π-electrons in 3-D. In a planar molecule each of the π-orbital and POAVs are perpendicular to the molecular plane, or to the three σ-bonds. Orbital orthogonality of π-orbital in one, two, and three dimensions is presented in Figure 4.2-1. The unique angle, $\theta_{\sigma\pi}$, is equal to 90° in a two-dimensional system (planar). For

F:4.2-1

π-ELECTRONS

(In three dimensions)

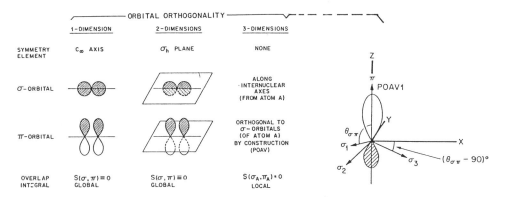

Orbital orthogonality of π-orbital in one, two, and three dimensions. This is a basis for the definition of a π-orbital and for the separation of σ- and π-orbitals which is applicable in all (1-3) dimensions.

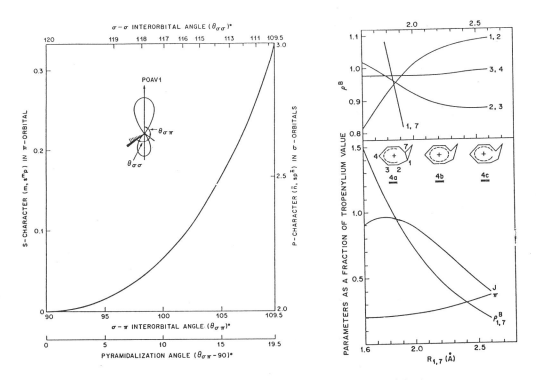

Relationship between the σ-σ and σ-π interorbital angles and the hybridization at a carbon atom between the extremes of sp^2 (planar geometry) and sp^3 hybridization (tetrahedral geometry) in C_3, symmetry (Left). Three-dimensional Hückel molecular orbitals analysis of the calculated geometries where π-electrons are the fractional contribution of the overlap integral.

nonplanar conjugated carbon atom, as in C_{60}, a simple approximation was used; π-orbital makes equal angles to the three σ-bonds in nonplanar geometries. This leads to definiton of a pyramidalization angle ($\theta_{\sigma\pi}$ - 90) which provides a convenient index of the degree of nonplanarity at a conjugated carbon atom. An important principle was found: conservation of orbital orthogonality in non-planar conjugated organic molecules. The Hückel molecular orbital (HMO) theory was extended into three dimensions through the resonance integrals (β) in standard HMO theory using the overlap integrals (from POAV π-orbital basis set). The π-orbital overlap integral between two atoms (k and m) is:

$$S_{km} = (s^k \negthinspace,\negthinspace s^m) + (s^k \negthinspace,\negthinspace p_\sigma^m) + (s^k \negthinspace,\negthinspace p_\sigma^m) + \\ (p_\sigma^k \negthinspace,\negthinspace p_\sigma^m) + (P_\pi^k \negthinspace,\negthinspace p_\pi^m)$$

(4.2.1)

and gives HMO-3D resonance integral

$$\beta_{km} = (\frac{S_{ij}}{S})\beta_o$$

(4.2.2)

where: S is the reference overlap integral and β_o the standard resonance integral ($\beta_o = H_{km} - ES_{km}$, where H_{km} is the resonance integral at secular determinant, and E is energy).

From the standpoint of π-bonded structure, electron-phonon coupling on the electronic structure and bond lengths of the C_{60} molecule are interesting. This approach leads to excited states which involve localized lattice distortion, including *solitons* and *polarons*. Hayden and Mele (1987) studied both the neutral state and effects of electron-phonon coupling "on the states for which one electron is added, one electron is removed on the first neutral excited state with one electron and one hole." They found that the neutral ground state has icosahedral symmetry, while in the new state (the result of electron-phonon coupling), degenerative states are involved. A degenerative state, known in literature as the Jahn-Teller effect, plays a very important role in the energy state of the C_{60} molecule and leads to vibration-rotation energy spectra.

4.3 Jahn-Teller Effect in Icosahedron Molecules

While in the Institute of Niels Bohr at Copenhagen in 1934, Teller and Landau talked about degenerative electronic states in the linear CO_2 molecule. The idea of Teller and his student, Renner, was that "an intricate coupling between the splitting of electron states and nuclear vibrations will arise which will modify the applicability of the Born-Oppenheimer approximation to these states." The idea, at that time, was too advanced to be acceptable, and Landau objected. As usual, great authority may slow down a new idea, but if the idea is right, sooner or later it will win acceptance. It was the right idea in this case, and Teller, with Jahn, won. As a result, the Jahn-Teller effect is well-known today: the energy in nonlinear molecules is minimized when distortions occur to remove either spin or orbit degeneracy, or both (Jahn-Teller, 1937). This is true for all nonlinear molecular systems except for those systems which have a spin with twofold degeneracy of the Kremers type.

Energy (Hückel-type approximation) for any conjugated molecule as a function of the bond lengths (a_{km}) is:

$$E = \sum f_\sigma(a_{km}) + 2\sum b_{km} \beta(a_{km}) \qquad (4.3.1)$$

where: $f_\sigma(a_{km})$ is the contribution of the σ-electrons to the bond between the "k" and "m" carbon atoms, b_{km} is the bond orders, and β is the resonance integral between the π-orbitals based on the "k" and "m" carbon atoms.

The resonance integral can be written in the form:

$$\beta(a) = \beta_0 \exp[\frac{-(a-1.4)}{c}] \qquad (4.3.2)$$

where: β_0 is the standard resonance integral ($\beta_0 = H_{km} - ES_{km}$) and "c" is a constant in the same unit of length as the distance between the "k" and "m" atoms.

For the first approximation we usually consider energy contributions (electronic, vibration and rotation) separately. Electronic energy transitions are usually in the electromagnetic spectrum of the ultraviolet and/or visible regions. Molecular vibrations are in the infrared region, while pure rotation is in the microwave region or the far infrared. By considering *electronic spectra*, a rough "picture" of the molecule is obtained. We will have a better "picture" of the molecule if we consider *electronic*, *vibrational* and *rotational* spectra together (Figure 4.3-1). Different theoretical approaches give different results ("pictures" of the C_{60} molecule) because researchers use different energy field models. In this book, we propose and present the HWK (Harter-Weeks-Koruga) model as the theoretical-experimental equivalent. Independently, Harter-Weeks developed the equivalent theoretical rotation-vibration spectra of icosahedral molecules (Harter and Weeks, 1989; Weeks and Harter, 1989; and Harter and Reimer, 1991) as Koruga's experimental-theoretical model of C_{60} surface energy (SE) based on STM image (Figures 4.1.2-1, 4.1.2-6 and 4.1.2-7). An experimental result, based on an STM image, showed that the optimal bias voltage to image C_{60} with internal structure was 0.0201 V. This value is much lower than expected (about 1 V). A second experimental surprise was that an electron may tunnel into and out of the C_{60} molecule (positon 10 in Figure 4.1.2-4 as a representation of two atoms from opposite side of the C_{60} molecule). Based on STM *icosahedron images* of the C_{60} molecule and energy state near the Fermi level, Koruga has developed the first step of a theoretical model based on a crystal field of icosahedral symmetry using group-theoretical and tensor-operator methods. During his research, Koruga has found that the Harter-Weeks work in this field are complementary to his own, although the history of the HWK model and the picture of the C_{60} molecule go back to the 1960s.

One year before the Kroto/Smalley research team (1984) identified carbon clusters of 60 atoms as icosahedral, Jacques Raynal (Raynal, 1984) published a purely theoretical paper about icosahedral groups and quantum harmonics along an axis of order five. He was inspired by a McLellan paper (1961) about icosahedral harmonics, whose main result was an application "to calculate the anisotropic magnetic splitting factors for the doublet states arising from the above symmetry." A crystal field of icosahedral symmetry was a subject of Judd's research using group-theoretical and tensor-operator methods (Judd, 1957), which was also Koruga's starting point in explaining the STM image of the C_{60} molecule.

The icosahedral group has five irreps A, T_1, T_2 (or T_3), G and H of

F:4.3-1

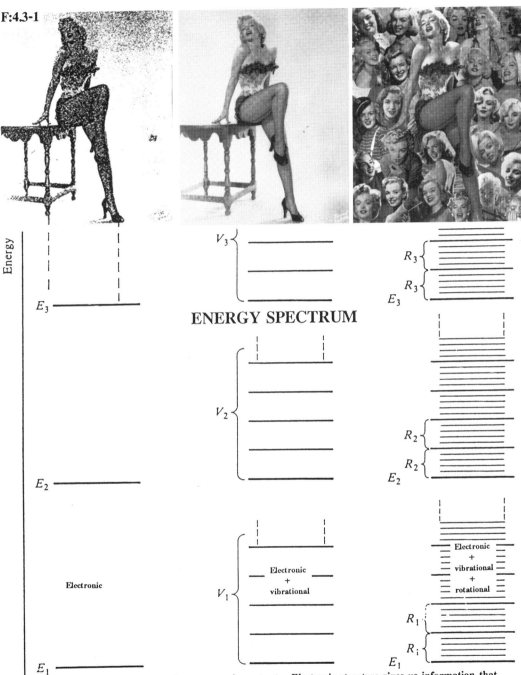

Energy spectra of Fullerene C_{60} are very important. *Electronic structure* gives us information that allows us to recognize C_{60} as a physical object and some of its properties. *Electronic and vibrational spectra* gives us more information about it, while *electronic, vibrational and rotational spectra* give us fine structure, so that we can recognize C_{60} by different rules. Analogous with Marilyn Monroes's figure: In the first picture, we can recognize Marilyn; her image in the second picture is better than in the first one, but we can also see that there is something in the background; the third picture gives us more complete information about her character, because it contains all of her faces from her acting roles.

ROTATION-VIBRATION SPECTRA OF C_{60}

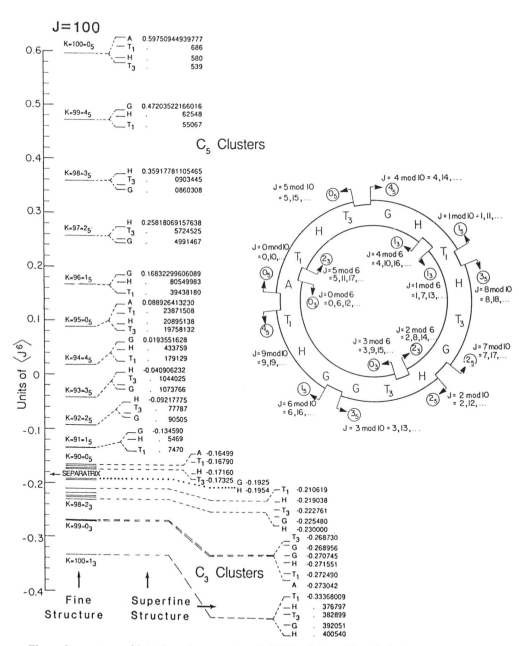

Eigenvalue spectrum with total angular momentum J = 100 for sixth-rank icosahedral tensor Hamiltonian $T^{[6]}$ Sequencing ring of spectrum is presented for arbitrary angular momentum J.

dimension $d^\alpha=1$, 2, 3, 4 and 5, respectively. The number d^α gives the degeneracy of the orbitals ("icosaharmonics") that "would result in a quasi-crystal-field splitting of ordinary J (*angular momentum*) orbitals or spherical harmonics Y_M^J belonging to a J level of degeneracy $2J+1$" (Harter and Weeks, 1989). The invariant sixth-rank icosahedral tensor has the form:

$$T^{[6]} = aT_0^6 + b(T_5^6 - T_{-5}^6) \tag{4.3.3}$$

where: $a=\sqrt{11/5}$ and $b=\sqrt{7/5}$. The tenth-rank invariant icosahedral tensor can be written as:

$$T^{[10]} = cT_0^{10} - d(T_5^{10} - T_{-5}^{10}) + e(T_{10}^{10} + T_{-10}^{10}) \tag{4.3.4}$$

where:

$$c = \frac{\sqrt{3 \cdot 13 \cdot 19}}{75}, \ d = \frac{\sqrt{11 \cdot 19}}{25} \ and \ e = \frac{\sqrt{3 \cdot 11 \cdot 17}}{75}. \tag{4.3.5}$$

In this case, the molecular Hamiltonian will be composed of both scalar(S) and tensor(T) parts in the form:

$$H = H_S + H_T \tag{4.3.6}$$

where the scalar part is

$$H_S = \upsilon_{vib} + BJ^2 + DJ^4 + ... - 2BJ \cdot 1. \tag{4.3.7}$$

Notations in Equation 4.3.7 have the following meanings: υ_{vib} is vibrational energy, BJ^2 is the rigid spherical rotor energy, DJ^4 is an isotropic distortion energy, while $-2BJ \cdot 1$ is the first-order isotropic Coriolis interaction. The scalar part of Hamiltonian gives us the excited state levels for different N for each J ($J-1$, J, $J+1$) but cannot split the $(2N+1)$-fold degeneracy of the angular momentum levels.

The second part of the molecular Hamiltonian, the tensor part, which is responsible for rotation level fine-structure splitting can be written in form:

F:4.3-3   **SUPERFINE STRUCTURE OF C$_{60}$ CLUSTERS**
                                   Local symmetry approach

a) C$_5$ Clusters

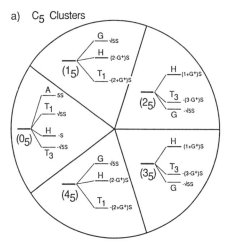

In fivefold clusters a new basis is formed by the projection of an initial state $|1, n_5\rangle$ with the icosahedral projection. Only the first row of the initial tunneling matrix is required. The cluster center-of-gravity energy has been set to zero and superfine splitting has been expressed in terms of nearest-neighbor tunneling amplitude.

b) C$_3$ Clusters

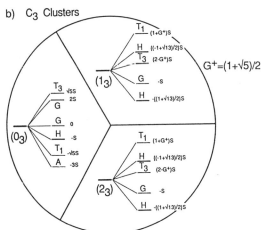

$G^+=(1+\sqrt{5})/2$

The C$_3$ clusters have the same fashion as the C$_5$ ones. However, there are two icosahedral G irreps in O$_3$, two H irreps in the 1$_3$ and 2$_3$ columns. Only nearest-neighbor tunneling is considered.

c) C$_2$ Clusters

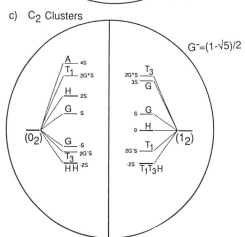

$G^-=(1-\sqrt{5})/2$

For C$_2$ clusters O$_2$ and 1$_2$ induced representations, there will be one three-dimensional and several two-dimensional blocks. Only nearest-neighbor tunneling is considered.

F:4.3-4 ROTATIONAL ENERGY SURFACES OF C$_{60}$
Mixed Sixth- and Tenth-Rank Icosahedral Tensor Hamiltonians

$$H(\vartheta) = B|J|^2 + (\cos \vartheta)T^{[6]} + (\sin \vartheta)T^{[10]}$$

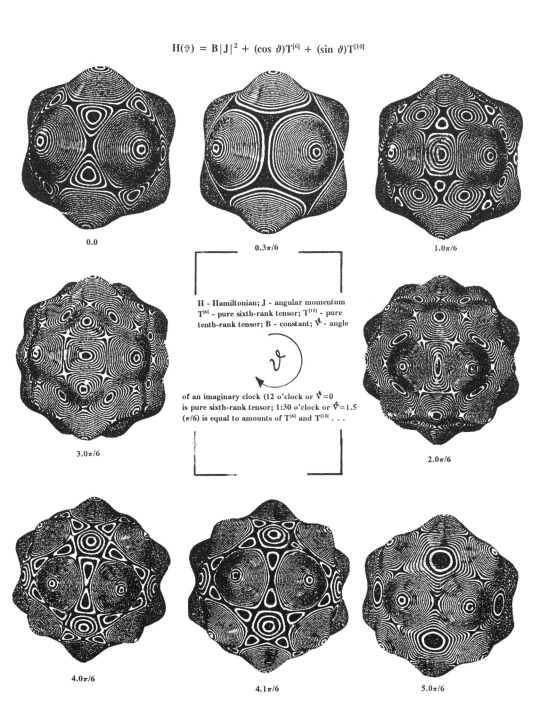

0.0

0.3π/6

1.0π/6

H - Hamiltonian; J - angular momentum
T$^{[6]}$ - pure sixth-rank tensor; T$^{[10]}$ - pure
tenth-rank tensor; B - constant; ϑ - angle

ϑ

of an imaginary clock (12 o'clock or ϑ=0
is pure sixth-rank tensor; 1:30 o'clock or ϑ=1.5
(π/6) is equal to amounts of T$^{[6]}$ and T$^{[10]}$ . . .

3.0π/6

2.0π/6

4.0π/6

4.1π/6

5.0π/6

$$H_T = \sum_{r_1 r_2} (t_c^6 T^{[6]}[r_1, r_2] + t_c^{10} T^{[10]}[r_1, r_2] + ...) \qquad (4.3.8)$$

where: t_c^6 is the molecular tensor constant (in Harter-Weeks notation $t_{r1, r26}$) and $T^{[6]}$ and $T^{[10]}$ are icosahedral tensors from Equations 4.3.3 and 4.3.4.

The simplest tensor Hamiltonian which consists of both scalar (H_S) and tensor (H_T) has the form:

$$H = H_S + H_T = BJ^2 + t_{066} T^{[6]} \qquad (4.3.9)$$

for which fine structure of eigenvalues is given in Figure 4.3-2. For the values $|J| = 1$, $B=1$ and $t_{066} = 0.2$, the surface of the rotational energy of the icosahedral structural is the same (or similar) as in Figure 4.1.2-6 (STM image and Koruga's model of C_{60} based on image reconstruction).

Quantitative values of superfine structure based on a local symmetry approach are given in Figure 4.3-3. G irreps, which are 4-dimensional, have the value of the Golden Mean ($G^+ = (1+\sqrt{5})/2$ and $G^- = (1-\sqrt{5})/2$) and are represented in all three clusters C_2, C_3 and C_5.

The eigensolutions of a mixed-tensor operator can be written in the form

$$H(\upsilon) = B|J|^2 + (\cos \upsilon)T^{[6]} + (\sin \upsilon)T^{[10]} \qquad (4.3.10)$$

where υ is artificially used angle (as an imaginary clock) and presented in Figure 4.3-4. For example, for $\upsilon = 0$ (12:00, as a starting point), only pure sixth-rank tensor ($T^{[6]}$) exists, for $\upsilon = 3$ ($\pi/6$) it is pure tenth-rank tensor ($T^{[10]}$) and we can say that it is 3 o'clock. When $\upsilon = \pi$, it will be 6 o'clock and pure sixth-rank tensor will be negative ($-T^{[6]}$) from Equation 4.3.3, and so on.

This approach shows that icosahedral structure, like that of C_{60}, may possess its own clock and be used as for building spatio-temporal molecular control devices.

PART II

FROM NANOBIOLOGY
TO NANOTECHNOLOGY

*Universe is technology--the most
comprehensively complex technology.
Human organisms are Universe's most
complex local technologies.*

– Buckminster Fuller

CHAPTER 5

BIO-FULLERENES

Fullerenes offer capabilities for nonoscale devices which can perform functions normally implemented by biomolecular assemblies. This chapter describes several of those assemblies: clathrin; microtubules; and centrioles.

5.1 Clathrin as a Bio-Fullerene

Clathrin is a Fullerene-like protein with a molecular weight of 180,000 daltons (dalton: "D" is the equivalent of the weight of one hydrogen atom). Clathrins are the major components of coated vesicles, important organelles for intracellular material transfer including synaptic neurotransmitter release. Coated vesicles have been observed by electron microscopy in many different cells (Pearse, 1976), and they consist of two main parts: clathrin and assembly proteins. There are three types of assembly proteins (AP)--AP-1, AP-2 and AP-3 (Keen, 1987)--which are composed of protein dimers. AP-1 has principal components with molecular weights of 108,000, 100,000, 47,000 and 19,000 daltons, while AP-2 contains approximately one molecule each of the 100,000, 50,000, and 16,000 molecular weight polypeptides per clathrin trimer. The third form of assembly protein, AP-3, contains 155,000 and 100,000 molecular weight polypeptides. Perhaps the most interesting assembly protein complex is AP-2, because it contains two subunits of about 50,000 daltons. Based on molecular weights, isoelectric points and antigenic determinants, two proteins, α- and β-tubulin subunits, have been found to be associated with coated vesicles in both bovine brain and chicken liver (Kelly et al, 1983).

The basic form of clathrin is a *trimer* (three subunits), but the basic form of assembly protein which associates with clathrin is a *dimer*. Clathrin is highly conserved in evolution and is composed of three large polypeptide chains and three smaller polypeptide chains that together form a *triskelion*. A different number of *triskelions* assemble into a basket-like network of 12 pentagons and various numbers of hexagons, **which is identical to the topological structure of Fullerenes!**

Clathrin triskelions are the assembly units of the sphere lattice on the surface of coated pits and coated vesicles. Assembly proteins (AP-1, AP-2, AP-3) have the ability to stimulate clathrin assembly under suitable physiological conditions of pH, ionic concentration and near-absence of divalent cations. Initiation for assembly comes from receptor-mediated endocytosis, ATPase and CV-associated kinases. The process of assembly/disassembly is very complex and poorly understood (Schmid, 1992). What is known is the structure of CV and clathrin. Neural cells (neurons) contain CV and clathrin with 12 pentagons and 20 hexagons with diameters of 70-80 nm. Most neuronal CV and clathrin are concentrated in synapses, some associated with microtubules (Dustin, 1984). Usually there are five CV per one microtubule, which through other cytoskeletal proteins contacts the presynaptic membrane.

BIOFULLERENE - CLATHRIN
                              Self-assembly

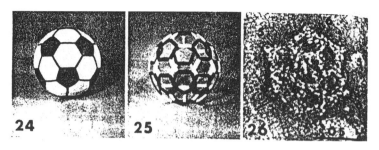

Kanaseki and Kadota discovered clathrin and proposed its icosahedral
symmetry in 1969.

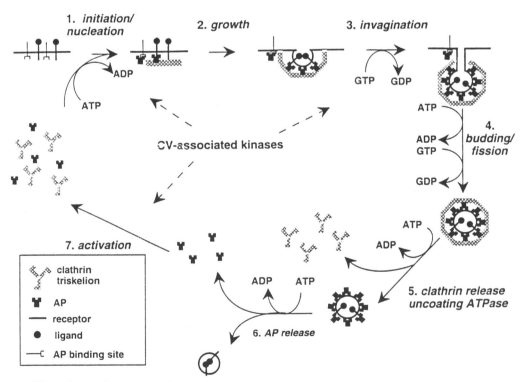

The dynamic cycle of coat assembly/disassembly which drives receptor-
mediated endocytosis. Reactions 1 through 7 constitute a model for the
dynamic cycle of coat assembly and disassembly which is coupled to receptor-
mediated endocytosis.

The inside space of CV may be occupied or empty; experimentally, both situations have been observed (Heuser and Firchhausen, 1985). The structure of CV and clathrin may be explained mathematically through a 3-D structure containing singly connected surfaces. Let V be the number of vertices, E the number of edges, and F the number of faces enclosed by the edges. The number of edges meeting at a vertex may be notated as r: the valency of the vertex toward the edges. The face has an equal number of vertices and edges, so the edge valency of a face (which we will notate as n) is equal to the vertex valency of that face. Fuller also used these notations for calculation and design of his geodesic domes.   The starting point for this calculation is Euler's relationship for polyhedra or two-dimensional closed figures. If the polyhedron is to have at most two different edge valencies among its vertices, then r_a is the number of one of them while r_b is the number of the other. Based on Euler's relationship, the number of vertices with "a" valency is:

$$V_a = 2r_b + [(1/2n - 1)r_b - n]F/r_b - r_a \qquad (5.1)$$

while the number of vertices with "b" valency is:

$$V_b = 2r_b + [(1/2n - 1)r_a - n]F/r_a - r_b. \qquad (5.2)$$

If we have triangular faces ($n=3$) and 5-valent ($ra=5$) and 6-valent ($rb=6$) vertices, then for $V_5=12$, there are the number of faces of 6-valent vertices (hexagons) which may be incorporated into the structure. If in Equations 5.1 and 5.2 valency of faces (n) and edges meeting at a vertex (r) exchange places, then vertices (V) and faces (F) also have to exchange places, and the 3-D figure becomes one with pentagon and hexagon faces.   This is a truncated icosahedron. Clathrin as a truncated icosahedron (same symmetry as C_{60}) is dominant in neurons, while clathrin with 30 hexagons is found in other types of cells (e.g. liver, etc.). Large bio-Fullerenes of 60 hexagons with a diameter of about 120 [nm] have been found in fibroblasts. The number of amino acids with negative charges (aspartic acid and glutamic acid) is 378 per clathrin, while the number of amino acids with positive charges (histidine, lysine and arginine) is 201. This means that there are 77 more electrons than positive

F:5.1-2   **CLATHRIN**

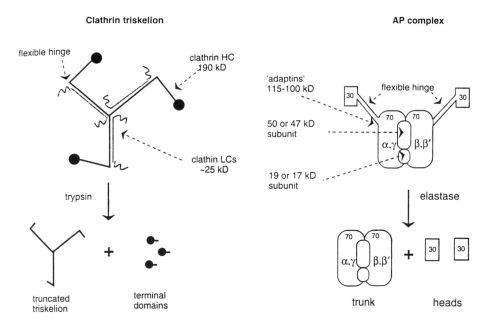

Summary of the structure of clathrin and AP complexes, the major coat constituents of clathrin-coated vesicles.

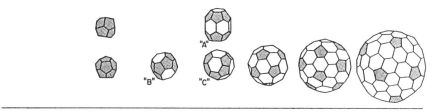

	Dodecahe-dron	Type "B"	Types "A" and "C"	Truncated icosahedron (brain)	(Liver)	Full icosahedron (fibroblasts)
No. of hexagons	0	4	8	20	30	60
No. of vertices	20	28	36	60	82	140
Predicted diameter (nm)	40	50	60	76	90	120
Predicted weight (MD)	13	18	23	38	53	77
No. of "facets" in deep etch	3–4	4–5	7–8	10	14	24
No. of "spikes" in neg. stain	4–5	5–6	6–8	8	12	16–17

Schematic drawings of several symmetrical basket designs of progressively larger sizes. The table below catalogues their expected complement of polygons, expressed as a number of hexagons in addition to the 12 pentagons always needed to form a closed polygonal shell (pentagons are shaded in the drawings).

F:5.1-3   **CLATHRIN VARIATIONS**

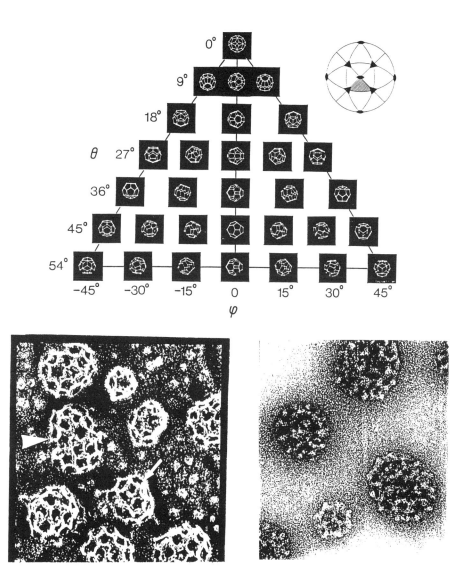

Gallery of views of structure B, covering the range indicated by the hatched area of
the stereogram. The symmetry axes marked correspond to the 23 subgroup of **43***m*
point group of the polyhedron. The full range of views can be filled in by using the
diagonal mirror lines passing through the 3-fold axes and then the rotation axes
marked. It should be noted that the line $\varphi = 0°$ is an anti-mirror for the gallery of
views. That is to say the view at $(\theta, -\varphi)$ is the same as that at (θ, φ) but reflected in
a vertical mirror and rotated 180°.

charges per clathrin molecule. Per "cage" there are about 8.316 more electrons than positive charges. This indicates that in the clathrin lattice, electrical current should flow through pentagon and hexagon rings. Also, similar to C_{60}, clathrin cages should rotate. However, there is yet no experimental evidence regarding electrical and magnetic properties of this clathrin.

Structures of CV and clathrin have a connection to a mathematical covering problem (Tarnai, 1991). It is shown that the problem of the covering of a sphere by circles may explain the different numbers of pentagons and hexagons in CV. If the number of circles on a sphere is 12, then CV possess pentagonal dodecahedron geometry. This kind of CV has been observed. Also observed have been $12_5 4_6$, $12_5 8_6$, $12_5 20_6$ and higher CV structures (12 pentagons and 4, 8, 20, etc. hexagons).

5.2 Clathrin-Microtubule Golden Mean Interaction

Microtubules (MT) are cylindrical protein polymers present in nearly all eukaryotic cells. They are composed of equimolar amounts of the two globular 50,000-dalton subunits, α- and β-tubulin, each having a similar amino acid composition and a similar overall shape. The subunits of tubulin molecules are assembled into long tubular structures with an average exterior diameter of 25-30 nm, capable of changes of length by assembly or disassembly of their subunits. Assembly/disassembly is sensitive to cold, high hydrostatic pressure, several specific chemicals such as colchicine and vinblastine, and other factors. The stable form of tubulin both *in vivo* and *in vitro* is a dimer, and the functional unit in MT assembly is the α-β heterodimer.

Kelly et al (1983) showed that α- and β-tubulin subunits are molecular components of coated vesicles. Tubulin subunits, as α-β dimers, play a role in clathrin nucleation, aggregation, and assembly. This is one type of tubulin-clathrin interaction. When clathrin makes a cage as a 3-D icosahedron structure, CA interacts with microtubules (MT) in synapses. "In the giant axons of the lamprey (*Petromyzon marinus*) which contain up to 3000 MT, synaptic vesicles are closely associated with MT, about five vesicles being radially disposed around on MT" (Smith et al, 1970 and Dustin, 1984). Some experimental evidence indicates that the CV are hexagonally packed around a group of three MT. Clathrin as a 3-D structure has an icosahedron symmetry with normalized spherical representation of a pattern with divergence $d = \phi^3$, where $\phi = (\sqrt{5} + 1)/2$. For sphere $x_1^3 + y^3 + z^3 = r^3(\phi)$ and Fibonacci series representation of ϕ^3 on the surface of the sphere, there is one-to-one mapping

F:5.2-1   **BIOFULLERENE**
Clathrin-microtubule interactions in neuron

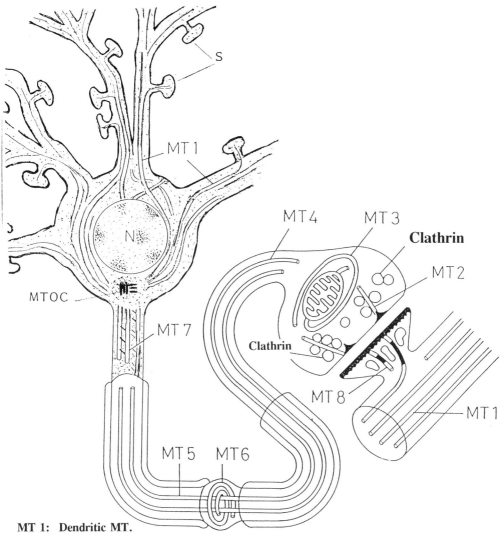

MT 1: Dendritic MT.
MT 2: Presynaptic MT, demonstrated by fixation in the presence of albumin, and contacting
       the synaptic membrane (VS synaptic vesicle).
MT 3: Helical MT surrounding a mitochondrion in a synaptic ending, mainly observed after
       incubation in Locke's liquid (artifact?, compared to the spiral MT of blood platelets).
MT 4: Axonal MT, in a non-myelinated region.
MT 5: Axonal MT in a myelinated axon.
MT 6: Helix of MT observed in Ranvier's nodes (perhaps artifactual).
MT 7: MT, close to the cell body, displaying thick intertubular "bridges."
MT 8: Post-synaptic MT.

MTOC: Microtubules Organizing Center
   N: Nucleus
   S: Spines

and we can write $\phi^3(r) = N(3)$. If clathrin-microtubule interaction is based on *Golden mean* (GM) principles, then the normalized cylindrical representation of a pattern of tubulin subunits of microtubules has to be with divergence $d=\phi^{-2}$, because $\phi^3 \cdot \phi^{-2}=\phi(\phi^N \cdot \phi^{1-N}=\phi^{0!})$ is eqvalent to $N(3) \cdot N(-2)_5=1$ (Figure 5-27) or $R_w^N \cdot S_w^{1-N}=1^{0!}$ (Equation 2.1.5.6). From the other side ϕ (golden mean) has the property: $1 + \phi = \phi^2$. As a double geometric ϕ-series, we can write: $1/\phi^2$, $1/\phi$, 1, ϕ, ϕ^2, ϕ^3. As we can see this is the only geometric series that is also a Fibonacci series, or in other words, *part* and *whole* have the same law. This double geometric ϕ-series has the properties: ... $1/\phi^2 + 1/\phi=1$; $1/\phi+1=\phi$; $1+\phi=\phi^2$; $\phi+\phi^2=\phi^3$ ... and shows what the link is between ϕ^{-2} and ϕ^3. This also indicates that structures which participate in synaptic information processing based on GM should have $1/\phi$, 1, ϕ and ϕ^2 properties. Here we will show that microtubules possess the property of ϕ^{-2}. From Chapter 2, it is clear that structures like clathrin which have icosahedron symmetry possess golden mean properties.

MT organization may be understood from the symmetrical patterns of spheres on a cylindrical surface. MT constitute a point set characterized by screw symmetry. For this, the basic symmetry operation *isometry* is the screw displacement (twist) as a product of rotation and translation parallel to the axis. This symmetry operation can transform point 0 to points: 1, 5, 13 . . .. Depending on screw displacements, this displacement gives ranks of points as arranged along helices--"parastichies." Parastichies are usually helices but in some cases can be circles or vertical generators of the cylinder. Cylindrical patterns of points may be constructed so that all points fall on a *single* generative helix (parastichy). Patterns with 2, 3, 4, . . ., k generative helices are bijugate, trijugate . . ., or k-jugate. For simplicity, a 1-jugate example will be presented.

Two parastichies, *m* and *n*, will be integers when a pattern is specified by citing two sets of parastichies. As shown in F:5.2-2 one might take $m = 5$ (0, 5, 10 . . .), $n = 8$ (0, 8, 16 . . .). As we can see, there is the possibility that another *m*-parastichy connects points 2, 7, 12, 17 . . ., or 2, $m+2$, $2m+2$, $3m+2$, ... Also for m-parastichy which contain the numbers: 1, 6, 11, 16, 21, . . ., we can write: 1, $m+1$, $2m+1$, $3m+1$, $4m+1$, ... To specify the symmetry properties of a pattern of points on a cylinder, it is necessary to define *angle of rotation* and *length of the translation*. One of the parameters which constitutes the screw displacement transforming point 0 to point 1 is the angular divergence $-\alpha$ (from $-\pi$ to π), while another is translational displacement $-h$ (dependent upon the radius of the cylinder, R). All other

F:5.2-2

MODEL OF SPHERES PACKING ON A CYLINDRICAL SURFACE

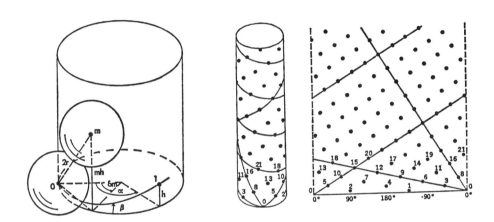

Packing unit spheres on the surface of a cylinder. Symbols: α-angular divergence, h-translational displacement, δm-secondary divergence, 2r-distance between centers of two tangent spheres, β-inclination of generative helix to the circumference of the cylinder (left). Each point represents sphere on a cylinder surface.

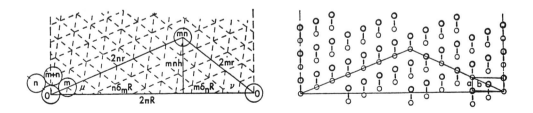

Diagram of hexagonal sphere packing on the surface of a cylinder. Symbols: μ-angle of inclination of m-parastichies and ν-inclination of n-parastichies.

F:5.2-3

TUBULAR SPHERES PACKING

Contacts $k(m, n, m+n)$	Divergence		Vertical displacement h	Radius of cylinder R	Inclination of parastichies		
	α (degrees)	δ_m (degrees)			μ (degrees)	ν (degrees)	ϕ (degrees)
Three strands							
3(0, 1, 1)	60.0	120.0	1.6329931	1.1547005	0.000	−53.481	53.481
(1, 2, 3)	131.810320	131.810320	0.6324555	1.0392305	14.818	−35.889	71.289
Four strands							
4(0, 1, 1)	45.0	90.0	1.6817928	1.4142136	0.0	−56.558	56.558
(1, 3, 4)	97.743120	97.743120	0.4693787	1.2905240	12.035	−43.116	69.617
2(1, 1, 2)	90.0	90.0	1.0	1.2247449	27.465	−27.465	90.0
Five strands							
5(0, 1, 1)	36.0	72.0	1.7013016	1.7013016	0.0	−57.858	57.858
(1, 4, 5)	77.414817	77.414817	0.3719648	1.5712212	9.938	−47.144	68.238
(2, 3, 5)	−141.843980	76.312040	0.3960824	1.4862538	21.810	−34.954	81.963
Six strands							
6(0, 1, 1)	30.0	60.0	1.7111994	2.0	0.0	−58.535	58.535
(1, 5, 6)	63.979611	63.979611	0.3075215	1.8651700	8.399	−49.688	67.156
2(1, 2, 3)	64.605543	64.605543	0.6513963	1.7692473	18.083	−39.716	77.684
3(1, 1, 2)	60.0	60.0	1.0	1.7320508	28.869	−28.869	90.0
Seven strands							
7(0, 1, 1)	25.714286	51.428572	1.7169464	2.3047649	0.0	−58.933	58.933
(1, 6, 7)	54.468269	54.468269	0.2618672	2.1663703	7.246	−51.385	66.304
(2, 5, 7)	−152.220650	55.558700	0.2759292	2.0623138	15.427	−43.000	74.923
(3, 4, 7)	−102.114200	53.657400	0.2845311	2.0037630	24.460	−33.885	84.782
Eight strands							
8(0, 1, 1)	22.50	45.0	1.7205911	2.6131259	0.0	−59.188	59.188
(1, 7, 8)	47.393551	47.393551	0.2278914	2.4719986	6.359	−52.623	65.622
2(1, 3, 4)	48.557790	48.557790	0.4781292	2.3615130	13.436	−45.394	72.950
(3, 5, 8)	135.960910	47.882730	0.2470401	2.2888251	21.179	−37.569	81.168
4(1, 1, 2)	45.0	45.0	1.0	2.2630334	29.363	−29.363	90.0
Nine strands							
9(0, 1, 1)	20.0	40.0	1.7230522	2.9238044	0.0	−59.361	59.361
(1, 8, 9)	41.931568	41.931568	0.2016509	2.7805347	5.659	−53.556	65.069
(2, 7, 9)	−158.479760	43.040480	0.2107296	2.6648441	11.889	−47.210	71.452
3(1, 2, 3)	42.948100	42.948100	0.6533239	2.5817903	18.654	−40.374	78.506
(4, 5, 9)	−79.658580	41.365680	0.2217071	2.5377417	25.828	−33.168	86.098
Ten strands							
10(0, 1, 1)	18.0	36.0	1.7247940	3.2360680	0.0	−59.484	59.484
(1, 9, 10)	37.590442	37.590442	0.1807896	3.0910831	5.094	−54.282	64.611
2(1, 4, 5)	38.604707	38.604707	0.3766097	2.9711125	10.654	−48.634	70.270
(3, 7, 10)	−107.068080	38.795760	0.1944493	2.8799837	16.654	−42.573	76.456
2(2, 3, 5)	−71.023387	37.953226	0.3970983	2.8223445	23.016	−36.178	83.092
5(1, 1, 2)	36.0	36.0	1.0	2.8025171	25.592	−25.592	90.0
Eleven strands							
11(0, 1, 1)	16.363636	32.727273	1.7260729	3.5494655	0.0	−59.574	59.574
(1, 10, 11)	34.058980	34.058980	0.1638156	3.4030824	4.630	−54.864	64.227
(2, 9, 11)	−162.513870	34.972260	0.1701343	3.2795451	9.647	−49.772	69.311
(3, 8, 11)	131.766390	35.299170	0.1754687	3.1818903	15.033	−44.340	74.821
(4, 7, 11)	98.721924	34.887696	0.1794356	3.1136676	20.735	−38.606	80.710
(5, 6, 11)	−65.269777	33.651115	0.1815468	3.0783987	26.659	−32.668	86.870
Twelve strands							
12(0, 1, 1)	15.0	30.0	1.7270401	3.8637033	0.0	−59.643	59.643
(1, 11, 12)	31.130975	31.130975	0.1497400	3.7161614	4.241	−55.339	63.901
2(1, 5, 6)	31.948959	31.948959	0.3102475	3.5896154	8.811	−50.715	68.517
3(1, 3, 4)	32.335872	32.335872	0.4794166	3.4865583	13.693	−45.790	73.485
4(1, 2, 3)	32.180692	32.180692	0.6539269	3.4098054	18.852	−40.602	78.775
(5, 7, 12)	150.281250	31.406250	0.1658479	3.3622191	24.225	−35.213	84.317
6(1, 1, 2)	30.0	30.0	1.0	3.3460652	29.717	−29.717	90.0
Thirteen strands							
13(0, 1, 1)	13.846154	27.692306	1.7277895	4.1785815	0.0	−59.696	59.696
(1, 12, 13)	28.664515	28.664515	0.1378816	4.0300657	3.912	−55.734	63.621
(2, 11, 13)	−165.301620	29.396760	0.1425214	3.9009549	8.105	−51.497	67.848
(3, 10, 13)	−110.065450	29.803650	0.1466362	3.7933449	12.568	−46.999	72.372
(4, 9, 13)	−82.549293	29.802828	0.1500103	3.7095399	17.274	−42.267	77.171
(5, 8, 13)	−138.133930	29.330350	0.1524231	3.6518984	22.179	−37.346	82.199
(6, 7, 13)	−55.273449	28.359306	0.1536854	3.6224690	27.216	−32.302	87.382

Parameters of models of hexagonally packed spheres of unit radius, whose centers are on a circular cylinder, classified by number of strands or near-longitudinal parastichies.

parameters can be derived from these: (1) Secondary divergence $\delta_m = m_a - 2t\pi$
and $\delta_n = n_a - 2t\pi$, where t is the integer which, in each case, gives $-\pi < \delta_m$
$\leq \pi$. The translational displacement is defined as: $h = R_\alpha \cdot tan\beta$, where R is
the radius of the cylinder. Numerical calculation for 20 different types of
sphere arrangements on a cylinder has been calculated by Erickson (1973) and
is presented in Figure F:5.2-2, 3, and 4. It was shown that tubulin subunits are
arranged on the cylinder by the same law of packing. Our intention here is to
show that tubulin subunit arrangements possess golden mean properties. We
consider the regular cylindrical representation of the tubulin subunits as a
growth of microtubules to the sequence 1, a, a+1, 2a+1, 3a+2, . . ., where
each term is the sum of the previous two. In Figure 5.2-5, the lattice of
tubulin subunits has a value of a=2, giving a divergence angle, as a horizontal
distance between consecutive points in the lattice, with a value of $[(\sqrt{5}+1)/2]^{-2}$
or ϕ^{-2}. In the case of ϕ^{-2} 13(8, 5), the divergence is d=137.51° with sequence:
1, 2, 3, 5, 8, 13, 21, 34, 55, 89, ... This is a picture of the structure of
microtubules as the normalized cylindrical representation of a tubulin subunit
pattern with divergence $d=\phi^{-2}$. Based on this knowledge, it is possible to
develop a dynamical model of clathrin-microtubule information processing
based on the golden mean, as a special *minimal entropy* model.

Generally speaking, *primary* activation of CV and clathrin in synapses
(together with other structures--neurotransmitters, . . .) correlates with neural
function and the awake state. CV and clathrin activate in our brain (neural
networks) ϕ^3 state, or information processing which produces a three-
dimensional information pattern capable of mapping one-to-one to the outside
world. In the case when microtubules and the cytoskeleton are *primary*
information processing elements in synapses (together with other structures) in
our brain (neural network), the ϕ^{-2} state is activated. This may correlate with
an unawake state, or sleep "in the world of shadow." Normally, both
processes exist at the same time, but one or the other is primary. We are in
the same *information state*, where both states exist at the same time. It is
similar to the situation of "Schrödinger's cat." Schrödinger described his cat
to illustrate a paradox in quantum mechanics: a real cat can be alive or dead,
but a quantum cat, hidden from view in a box, is neither alive nor dead. If
quantum mechanics plays an important role in biomolecular processes, then
clathrin and microtubules are good candidates for linking quantum mechanics
and biomolecular information processing. But the clathrin-microtubule
information processing model gives us an even more useful tool to explain
human creativity than does today's quantum mechanics. In Schrödinger's cat

F:5.2-4

TUBULAR ARRANGEMENTS OF SPHERES

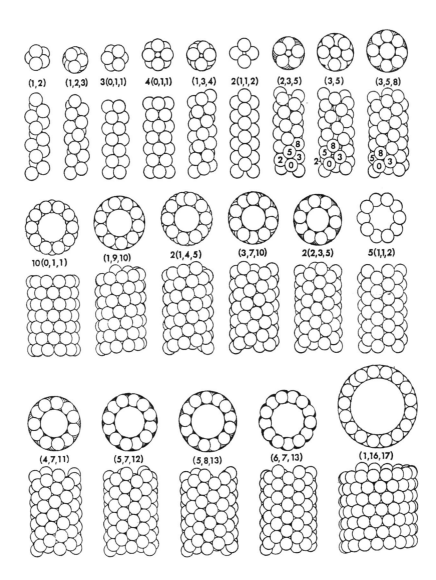

Parallel projection onto the plane of tubular arrangements of spheres. Parameters are concerned with screw displacements along m- and n-parastichies. When the distance from 0 to $m + n$ equals $2r$, the spheres are in contact along the m-, n- and $(m+n)$-parastichies, the pattern is a triple-contact or hexagonal packing pattern, and it may be designated (m, n, m+n), like (5, 8, 13).

model, one important event does not exist: the *immediate unity* of cat states in "neither alive nor dead." The information model of clathrin-microtubule opens possibilities for an immediate state of unity, N(3), and N(-2)$_5$ as a "world of shadow." According to this model, *immediate unity* of clathrin-microtubule information processing in synapses, based on Koruga's information complementarity (inversion operation of symmetry in information theory-- Equation 2.1.5a), gives a unity of real and shadow worlds: ***human creation***.

According to ancient Egyptian legend, **World** was *created* when the Sun God-Ra (N=5) unified with his *own shadow* (N=-4$_7$). This legend may parallel human information processing in synapses based on clathrin (N-3, icosahedron with golden property ϕ^3) and microtubule (N=(-2)$_5$, icosahedron with golden mean ϕ^{-2}--world of shadow) which unify to create the material world: music, art, cars, aircraft, ... We do this as God did (according to legend), by becoming one with our own shadow (which exists unto us). According to the bible, a human being looks like God (". . . the man is become as one of us. . ."--Genesis). Of course we cannot yet say whether these legends are true or are the products of imagination based on information processing in the human brain. But there is a very interesting common feature in these cases (our information model, Big Bang theory and legends): the starting point of *creation* is a *point*. In our model, N=0 is dimension "zero," with a value equal to one ("point"), and unity of information of pairs of dimensions (real and shadow), except N=1. According to Big Bang theory, the universe started from a *point* (some experiments regarding red shift and background radiation support this view). Legends say that World came from God and his interaction with his own shadow (ancient Egypt) or his word-- which is similar to shadow (Bible: ". . . in the beginning was God, and God was word . . ."). There is one more interesting legend: the Pleasgian Creation Myth, which says that ". . . in the beginning, there was only the Goddess Eurynome who rose nude from the Chaos, but not finding anything solid to lay her feet on, she separated the Waters for the Heavens, dancing her celestial dance on the waves. The dance gave birth to the great snake Ofion, which, not able to withstand the magic of Eurynome's dance, became one with her. Eurynome was then transformed into a pigeon, and, prostrate on the waves, laid the All-Embracing Egg. On her command, Ofion belted itself *seven* times around the Egg, heating it so that it burst, and all things of this world came to be: the sun, the moon, the stars, and the Earth, with all rivers, trees, plants and animals." Similar pictures of the *Creation* can be found in ancient Greece, China, India, ... In both, ancient (legends) and modern (physics--Big Bang)

F:5.2-5   **BIOFULLERENE**
                          Microtubules and golden mean

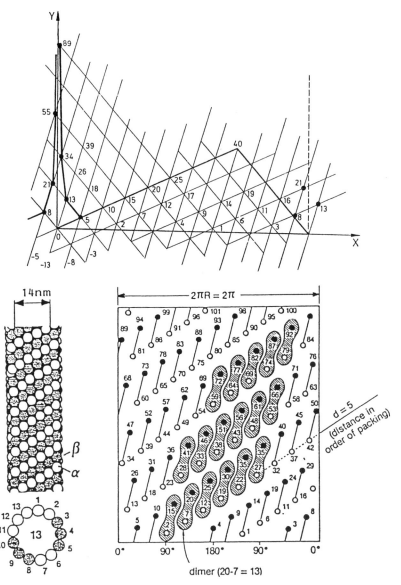

The golden ration $[(\sqrt{5} + 1)/2$ and $(\sqrt{5} - 1)/2]$ is the main property of the structure and energy of the C_{60} molecule. Also, the golden mean looks like an invariant of nature (similar to the speed of light) in the world of symmetry, because it gives optimal proportions between two parts of the same entity. If two parts are opposites like *plus* (+) and *minus* (-), the ratio of entity and/or process will express the symmetrical proportions involved in the generation of spiral patterns. Microtubules as a self-assembly device possess two spirals: α and β tubulin protofilaments on a cylindrical surface. The normalized cylindrical representation of a microtubule's pattern with divergence ϕ^{-2}, where $\phi = (\sqrt{5} + 1)/2$, is given in the picture. The Fibonacci numbers, on the polygonal lines, alternate on each side of the vertical axis due to the absence of intermediate convergence the continued fraction of ϕ^{-2}.

versions, *Creation* is a unit point.

Based on this consideration, the main question arises: Is a New Intelligent Being based on Fullerene C_{60} possible? If so, what will be its main characteristics? More about these considerations will be in Chapter 7.

5.3 Tubulin, Microtubule and Centriole

Tubulin is a globular (spherical) protein composed of amino acids. There are three general types of tubulin--α, β and δ--but only α and β participate in microtubules. α-tubulin subunits usually contain 450 amino acids, while β-subunits contain about 445 amino acids. There are about two times more negative charges (*Asp* and *Glu*) than positive charges (*Lys*, *Arg* and *His*). The diameter of tubulin subunits is about 4 nm. The *in vivo* and *in vitro* stable form of tubulin subunit is the α-β heterodimer, whose molecular weight is about 100,000 daltons. This heterodimer is the main building element of microtubules (Figure 5.3-1). Microtubules may exist from 7-17 protofilaments, but usually (85%) a microtubule contains 13 protofilaments.

Since some experimental results link tubulin and microtubules to bioinformation processes such as memory and learning (see Chapter 6), microtubules have become the subject of intensive research as bioinformation devices (Hameroff 1987, 1989; Koruga 1986, 1992). Based on the structure and dynamics of microtubules, sphere packing and code design (Leech and Sloane, 1971; Sloane, 1984), microtubule code systems were proposed (Koruga, 1986; 1990) which can explain why microtubules usually contain 13 protofilaments (Figure 5.3-2). From the aspect of structure, information in microtubules is stored in binary $K_1[13, 2^6, 5]$ code, while from the aspect of transmission, information in microtubules is stored in ternary $K_2[24, 3^4, 13]$ code. The relation between K_1 and K_2 codes, through digital sum, is given in Figure 5.3-1. For formation of a ternary K_2 code, microtubule-associated proteins (MAPs) are essential.

Lengths of microtubules may be from a few nanometers to hundreds of nanometers in cell cytoplasm, dendrites and synapses, while in axons they may be from 1 μ to 1 meter (motor neuron). Microtubules form networks within cells similar to networks of neurons in the brain. The connections between microtubules are through MAPs, which also may play a role in "switching" processes (Figure 5.3-3).

Cell structure is organized from a central focal region near the nucleus

MICROTUBULES
                    Binary and terniary code systems

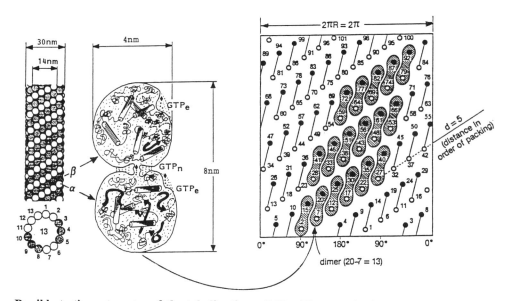

Possible tertiary strucutre of the tubulin dimer (left). The organizational form of tubulin in microtubules (right). Microtubules in unrolled form. The position of a dimer is visible in the row of packing during the process of polymerization. Each dimer carries the value of the number of protofilaments in a microtubule, making the "information power" of the dimer and microtubule the same.

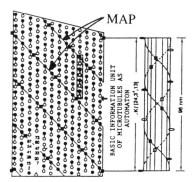

Digital Sum	CODE	
	$K_1[13,2^6,5]$	$K_2[24,3^4,13]$
0	18	19
1	32	16^+ 16^-
2	10	10^+ 10^-
3	4	4^+ 4^-
4	–	1^+ 1^-
All	$2^6 = 64$	$3^4 = 81$

From the aspect of structure, information is stored in the K_1 code. However during the transmission of information the binary K_1 code is transferred into the ternary $K_2 = K_1$ code. A conformational change occurs as hexagonal organization of subunits changes to cubic. In order for this change to have sense and for mapping from K_1 to K_2 to occur, there has to exist a unit by which this change occurs. This case is a cell 12 x 13 dimers, ordained by the basic elements of microtubule codes K_1 and K_2. Eighteen binary words of the K_1 code yield 19 tertiary words of K_2 code; 32, 10 and 4 from K_1 code yields as symmetric and asymmetric in K_2 code. Finally, two more "nonsense" ("stop") words and ambiguities are seen in the coding of amino acids in the genetic code.

F:5.3-2   **SPHERE PACKING AS CODE DESIGN**

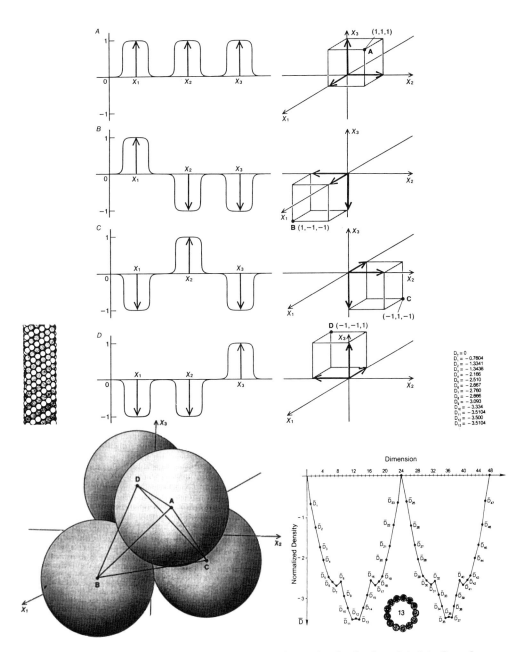

Design of a code for the efficient transmission of information is closely related to the sphere-packing problem. The code is to be a finite set of signals called code words that are easily distinguished from one another and do not waste electric power. If each code word is a sequence of, say, three discrete voltage levels, each sequence can be represented as a point in three-dimensional space: the first coordinate of the point is the numerical value of the first voltage level, the second coordinate is the value of the second voltage level and so on. Packing of the spheres gives that the optimal number is 11, 13, 35 and 37.

F:5.3-3   MICROTUBULE NETWORKS WITHIN CELLS

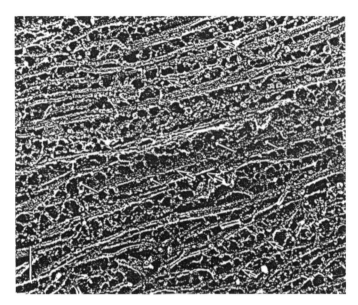

Microtubules with other cytoskeleton elements make a network in the cell (cell "brain") in a way similar to that in which neurons make a network in the brain.

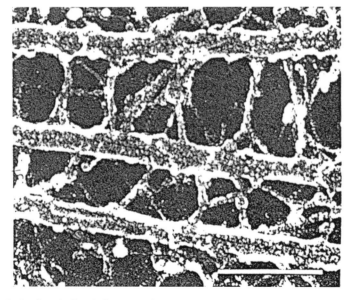

Three microtubules in cell "brain" and their connection through MAP (microtubule associated proteins).

called the microtubule organizing center (MTOC). The principle component of this center is the centriole, an organelle which consists of 2 perpendicular cylinders. Each of these cylinders, about 400 nm in length, is made up of nine MT triplets. The triplets are formed of one complete MT, with 13 protofilaments, and second and third partial MT with 10 protofilaments. Three protofilaments are shared between the first and second, and second and third MT. The triplets are held together by MAP-like links which extend between triplets. At the base each triplet is attached to a central core of dense material known as a "cart-wheel" (Figure 5.3-4 and 5.3-5). Centrioles possess a nine-fold symmetry axis, and in each of nine points, there are three MT. The centriole has three main movements: (1) *rotation*, (2) *pulsing* and (3) *duplication* during cell division. Centrioles and the MTOC play key roles in dynamic coordination of cell cytoplasm and activities. When cells are separated into two parts (one with nucleus and cytoplasm, the other with centriole with some cytoplasm), the "new" cell with the nucleus cannot produce a new centriole (Figure 5.3-6) and cannot divide. Only cells which have both nucleus and centriole have normal biological properties of mitosis (cell division).

The centriole remains a central enigma in the cell and molecular biology. This enigmatic characterization may be resolved by considering centrioles as double Golden mean devices: first through microtubules (divergence ϕ^{-2}), and second through MT nine-fold symmetry triplets. These triplets (Fig. 5.3-7) may have left and right orientations with the golden mean angle (GMα). This angle is the result of synchronization of rotation, twisting and oscillatory processes of the centriole. This process is based on a circle map:

$$\theta_{n+1} = \theta_n + \Omega_n - \frac{k}{2\pi}sin(2\pi\theta_n) \qquad (5.3.1)$$

where: k is a "coupling constant," which regulates the degree of nonlinearity [for MT it is $1/2\pi$, as ratio $N(-2)_5/N(0)$], θ_n represents an angle in the phase space of a dynamic system, Ω is the bare winding number (representing a frequency ratio such as the resonant frequency of an oscillator based on Fibonacci series; for microtubules, it is F_{n-2}/F_n). This equation (5.3.1) is related with Farey-tree and through the "golden way" (1/2, 2/3, 3/5, 5/8, 8/13, 13/21 . . .) rapidly to *chaos*. Chaos means, in this case, that initially close values of θ will diverge exponentially so that all predictability is lost as the

CENTRIOLES
                               Nine-fold symmetry organelle

$$M(r) = \frac{3}{4\pi r^3} \int_o^{2\pi} d\Theta \int_o^{\pi} \sin\varphi \, d\varphi \int_o^r a \, a^2 \, dr = \frac{3}{4} r \, .$$

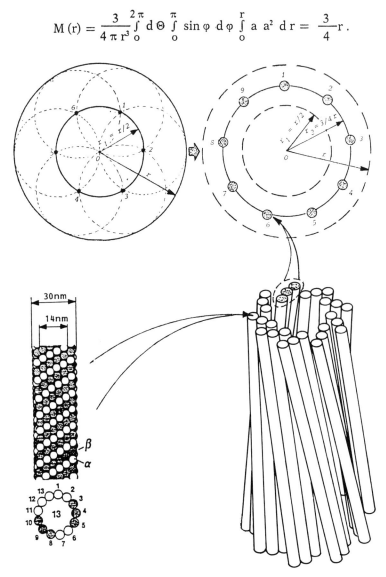

Nine-fold symmetry of centriole as a result of *wave-particle* synergy of electromagnetic field of tubulin subunit in pool (cell--non-polymerase form) and electromagnetic field of tubulin in microtubules (cytoplasm--polymerase form). Distribution value of tubulin subunits, M(r), shows that particle (microtubule) arrangement is 3/4 of r (bigger circle). The distance between microtubules on 3/4 r is the same as the distance between the basic circle $r_1 = r/2$. Oscillation (pulsing) of microtubules between r_1 and r is the result of twisting. Only tubulin subunit on middle of length of centriole always has value 3/4 r.

F:5.3-5   **CENTRIOLE**
                                Pulsing through twisting

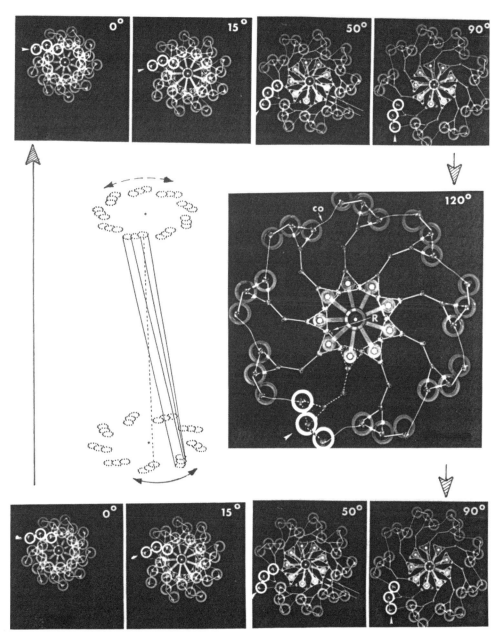

One end of microtubule pulsing in centriole from position 0° to 120° while the other one at the
same time goes from 120° to 0° position. There is one position on microtubules in centriole
where tubulin subunits are always in position 90°. This position is a link (through
electromagnetic field) with another centriole which is perpendicular. This model explains that
relationship between microtubules and centrioles, as part and whole, based on the 90° angle.

F:5.3-6

CENTRIOLE'S ROLES IN CELL

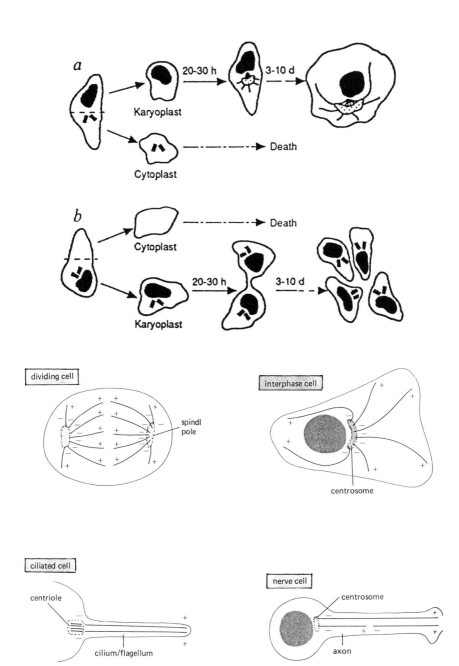

system evolves in time. In Farey-tree, organization of the rational numbers F_{n-2}/F_n (Fibonacci series) lies on a zigzag path approaching the golden mean or its square. In this case, F_{n-2}/F_n sequence beginning with 0/1 and 1/2 equals 1/3, 2/5, 3/8, 5/13, 8/21. The parameter Ω_n gives a dressed winding number equal to the frequency ratio F_{n-2}/F_n which may be determined numerically. For $\theta(0)=0$, $\theta(F_n)=F_{n-2}$ yields the following parameter values: $\Omega(1/2)=0.5$, $\Omega(1/3)=0.35166$, $\Omega(2/5)=0.40747$, $\Omega(3/8)=0.38826$, $\Omega(5/13)=0.39511 \ldots$ $\Omega_\infty=0.3933377$. These parameter values give rise to superstable orbits because the iterates θ_n include the value $\theta_n=0$ for which the derivative of the critical circle map vanishes. The golden mean of the centriole may be visible (as visual perception--F:2-9) through a calculation based on the equation:

$$(r_n, \phi_n) = (c \cdot r_{n-1}, \phi_{n-1} + \Delta\phi) \qquad (5.3.2)$$

where: c is a constant smaller than but arbitrarily close to $1/\sqrt{5}$, $\Delta\phi$ is angular increments (Figures 2-9 and 5.3-7).

As we can see (Figure 5.3-7), both experiment and theory show *left* and *right* orientation of microtubule triplets in centrioles. Based on Fuller's concept of synergy, this means that centrioles possess a unique role in cell dynamics, especially in our "shadow world." The supreme bio-Fullerene in terms of icosahedron structure is CV and clathrin, while the supreme bio-Fullerene in terms of "icosahedron shadow" (through golden mean as one of the main properties of the icosahedron) is the centriole.

The unity of *Rw* (real world) and *Sw* (shadow world) through the icosahedron may be realized through the C_{60} molecule. In biology, the unity of *Rw* and *Sw* is *indirect* (other structures participate in information processing between them), while in *Fullerene devices*, unity may be *immediate*. For this kind of unity, one of the conditions is that artificial microtubules and "centrioles" should be built of C_{60}. This unity of information compatibility based on icosahedron symmetry and golden mean property will be the ultimate goal of nanotechnology based on the C_{60} molecule.

CENTRIOLES
A central enigma of cell biology

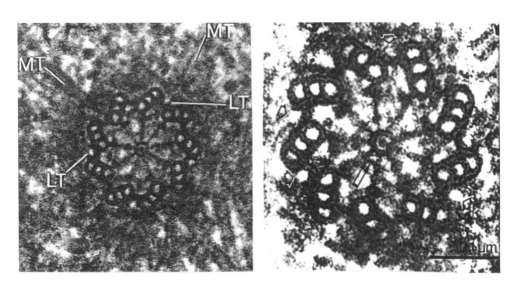

Centrioles with *left* and *right* orientation of microtubule triplet are found in cells.

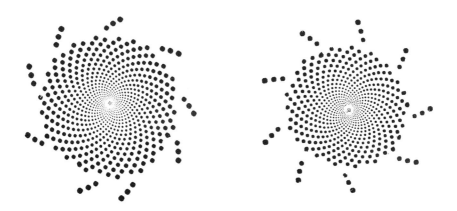

Left and *right* orientation of microtubule triplet in centriole as result of spatio-temporal oscillatory process of microtubules. This motion possesses golden mean properties, and through the Farey-tree organization (Fibonacci numbers-13) route to chaos.

F:5.3-8   **CENTRIOLE**

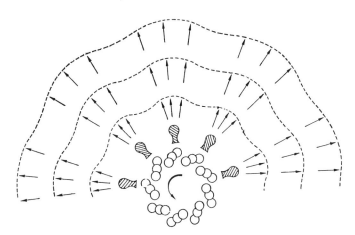

Schematic representation of the origin centriolar cylinder's role of oscillator for the propagation of phased signals in the cell. Centriolar cylinder's rotation would allow temporal control of this activity.

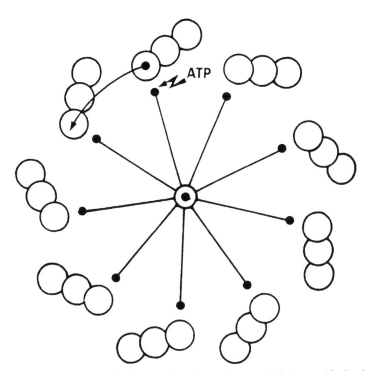

Schematic illustration of the mechanism for centriolar cylinder's rotation. The proximal end of a cylinder is represented. The spokes of the cartwheel are presumed to possess an ATPase activity at their peripheral extremities.

CHAPTER 6

INTELLIGENCE WITHIN CELLS: BIO-FULLERENE INFORMATION PROCESSING

F: 6 - 0   **SYMMETRY AND BIO-FULLERENES**

Stuart Hameroff, researcher in the field of molecular computing and information processing in microtubules (Bio-Fullerene) in front of his Advanced Biotechnology Lab at the University of Arizona, Tucson, USA.

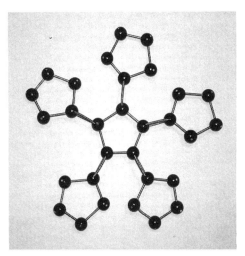 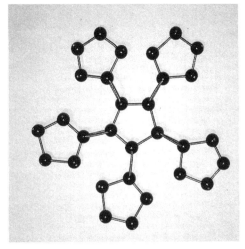

Pentagons cannot form 2-D lattice (in the sense of Bravais Lattice - inorganic crystals) but (left and right) max form a 3-D lattice on the surface of a sphere. This solution exists biologically within cells as clathrin, (Bio-Fullerene), and in laboratories today as human-made C_{60} molecule. Left-right symmetry is a crucial principle in biology (two sides of the body, DNA - proteins, triplet orientation in centriole, etc.) and will also be important in C_{60} nanotechnology.

C_{60} and other Fullerenes may in the future be utilized as nanoscale intelligent hardware. Beyond memory storage and processing elements, Fullerene structures may be capable of emulating intelligent hardware within living cells. This chapter reviews the function of such intelligent biomolecules.

6.1   Adaptive behavior

Most views of biological intelligence consider cells (neurons, immune cells, etc.) as fundamental units whose cooperative interaction yields collective, emergent properties.   In the brain/computer analogy, neurons and their synapses are likened to fundamental switches or gates. Neurons, however, are extremely complex and themselves resemble intelligent computers more than simple switches. The complexity and intelligence of single cells are also illustrated by the adaptive, "intelligent" activities of single cell organisms such as Paramecium.   Here we review biomolecular mechanisms of adaptive, intelligent behavior in 2 types of cells: Paramecium and brain cortical neurons.

6.1.1 Adaptive, intelligent behavior in Paramecium

In daily pond life, Paramecium samples regions of its environment by using its own cilia to create vortices of current. If the stimulus is favorable (i.e. bacterial food), the slipper shaped cell swims towards the food and begins to feed. If the stimulus is judged to be injurious, paramecia swim backwards by reversing the strokes of their cilia, turn toward one side, and swim forward in a new direction. Numerous stimuli can elicit such "reflex-like" responses to chemical, mechanical, pressure, fluid flows, gravity, temperature, electrical, and light stimulation (Wichterman, 1985). In the "rebounding reaction," a neutral obstacle is encountered and the organisms swim along it with their anterolateral side, then finally pass it swimming away at a new angle (Figure 6-1). Paramecium mating behavior involves associating with a partner, aligning and coupling of their single-cell bodies, and exchange of cytoplasmic "pronuclei" material (Figure 6-2). Other activities of paramecia have been ascribed to learning in that they may involve memory, conditioned reflexes, reinforcement, associations and behavior modification based on previous experience (e.g. Gelber, 1958).   For example, a number of studies have observed paramecia swimming and escaping from capillary tubes in which they could turn around. In general, results showed that with practice the ciliates

F:6-1 and 6-2

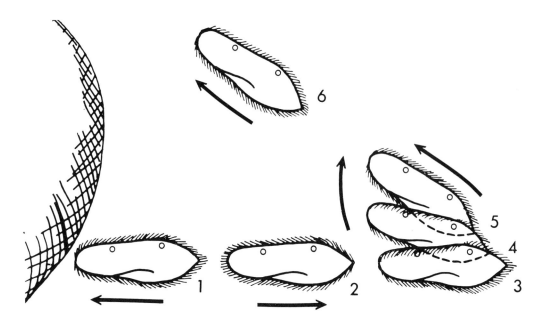

Rebounding reaction of paramecium towards external stimulus. (1-3) short-lasting ciliary reversal; (3-5) pivoting and circling; (6) swimming in new direction.

Mating behavior in paramecium. The single-cell organisms align with partners and exchange pro-nuclear genetic material.

took successively less and less time to escape, indicative of a learning mechanism (French, 1940; Applewhite and Gardner, 1973; Fukui and Asai, 1976). Many other experiments suggest paramecia can learn to swim in patterns and through mazes and have a short-term memory, although some of these behaviors depend on their environment (Applewhite, 1979). The learning and memory conclusions have been challenged (Dryl, 1974), but avoidance, rebounding, habituation, and attraction/avoidance responses are adaptive behaviors at least comparable to reflexes in multicellular organisms containing specialized nervous systems. How does Paramecium perform such complex tasks without benefit of a brain or a single synapse?

Adaptive behaviors in Paramecium involve observable motor functions which are performed by coordinated actions ("metachronal waves") of hundreds of cilia (Figure 6-3). Hair-like appendages comprised of 9 microtubule doublets arranged in a ring around a central microtubule pair, cilia are membrane-bound extensions of the cytoskeleton in cells ranging from protozoa to human epithelium. They may have sensory function in that their perturbation is transmitted to the cell, and several modes of ciliary motor movement can occur: non-cyclic (inclination) and cyclic (power stroke) bending, and changing rates of circular motion (beating frequency). These movements, utilized in avoidance, rebounding and attraction/avoidance behaviors by paramecia, as well as sweeping of mucus, dust and dead cells by human respiratory epithelial cells, are ascribed to contractile motor-like microtubule-associated proteins such as dynein which are attached at periodic intervals on ciliary microtubule doublets. These dyneins (powered by consumption of ATP: "ATPase") contract sequentially to effect a sliding of microtubule doublets along one another. Coherent sliding of the 9 doublets manifest functional ciliary movement. Membrane potentials (both bipolar and graded) in the ciliary and cell surface membranes are coupled to dynein ATPase activation by "second messenger" cyclic AMP and transmembrane calcium ion $[Ca^{++}]$ flux. This coupling, however, cannot explain numerous significant questions regarding the complexity of ciliary regulation. For example, how does $[Ca^{++}]$ act to induce 9 MT doublets to perform different sliding programs in sequence, and to perform the sequences differently at different levels of ciliary cross section? How can intraciliary $[Ca^{++}]$ reprogram doublet sliding to maintain polarity of the cycle, but with power stroke redirected? Perhaps most interestingly, how are the activities of hundreds of cilia orchestrated to perform complex pivoting, spinning and circling of the entire cell in response to external influences?

F:6-3 and 6-4

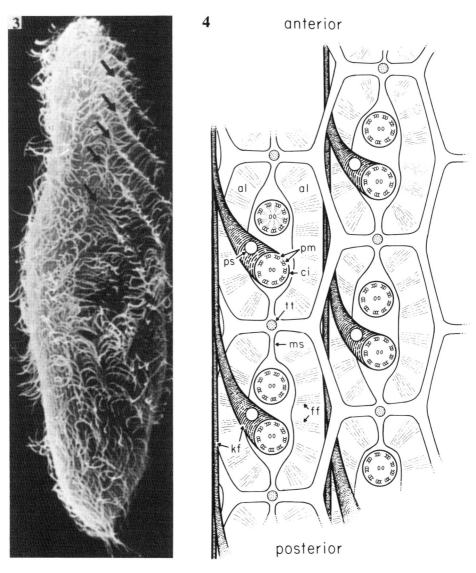

Scanning electron micrograph (rapid fixation) of Paramecium showing
protruding cilia. Arrows demonstrate ciliary movement in metachronal
waves.

Diagram showing paramecium polygonal infraciliary lattice with cilia
(ci), containing 9 + 2 sets of microtubules including peripheral
microtubule (pm) doublets.   ms: median septum, al: alveoli, tt:
trichocyst tips, ps: parasomal sacs, ff: fine fibrils, kf: kinetodesmal
fibrils. This illustrates how cilia (sensory and motor) are connected to
intra-cellular cytoskeleton.

F:6-5

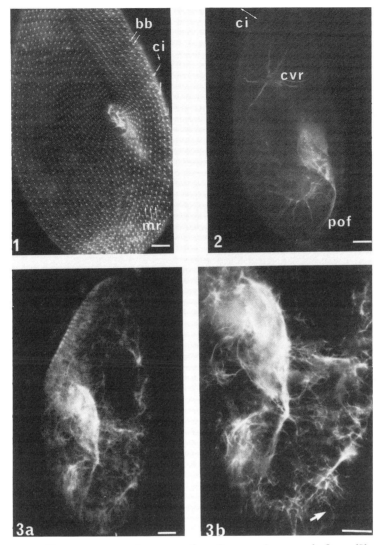

1.   Immunofluorescence image of sub-membrane infra-ciliary MT lattice in de-ciliated paramecium.   ci: few cilia remain despite de-ciliation; bb: basal bodies: cortical lattice attachments for cilia; mr: microtubule ribbons: scale bar: 10 μm (10^4 nm).
2.   Immunofluorescence image of acetylated ("post-translationally modified) α tubulin in paramecium interior. ci: cilia (poorly labelled); cvr: contractile vacuole roots; pof: post-oval fibers; scale bar: 10 μm.
3a &b.   Immunofluorescence images of internal microtubules in paramecium. (a) General view. (b) Enlargement of posterior part of same cell. Arrow: MT spanning from cortical lattice to internal system. Dense region of MT in center of posterior region includes MT organizing center, centriole and described as "neuromotorium"-- Paramecium's "central nervous system." Scale bar: 10 μm.

These phenomena may be explained by signaling, information processing and working memory in paramecia microtubules--both ciliary and intracellular. Atema (1974) proposed that signal transduction in sensory cilia involved propagating conformational changes along ciliary microtubule subunits, and Lund (1933) and Rees (1933) showed that MT within paramecia were "conductile," and transmitted information. (Arguments for propagative signaling, coherent conformational excitations and information processing in microtubules and microtubule networks will be presented later in this chapter.) Complex coordination of multiple ciliary beating ("metachronal patterns") are apparently coordinated from within the cell's cortical ectoplasm which contains cytoskeletal basal bodies and a hexagonal array of cytoskeletal fibrils ("infra-ciliary lattice") to which the cilia are anchored (Figure 6-4). This peripheral network is in turn interconnected to internal microtubules and other cytoskeletal structures. Several authors (e.g. Lund, 1933; 1941) identified a central region to which these cytoskeletal structures connect, and proposed that this centralized location was the focal point of integration of sensory input and control of motor response (Figure 6-5). This "neuromotorium" was seen by some authors as the "brain" of Paramecium; in essence it is a confluence of cytoskeletal structures. Thus the cytoskeleton may be viewed as the nervous system of the Paramecium, with signaling and information processing occurring via traveling conformational changes of microtubule polymer subunits. Metachronal patterns of ciliary beating are reversibly inhibited by the general anesthetic chloroform (Parducz, 1962), suggesting some functional link to cognition in higher organisms.

6.1.2 Adaptive, intelligent behavior in brain cortical neurons

External events which initiate adaptive responses via ciliary and/or membrane perturbation in Paramecium are similar to receptor binding of neurotransmitter molecules in neuronal synapses.   Each triggers second messenger cascades, which include G proteins, $[Ca^{++}]$, adenyl cyclase, cyclic AMP, membrane lipid breakdown products such as phosphatidylinositol, etc. These transducing elements, in turn, respond by activation of dynein ATPases and cytoplasmic kinases such as calcium-calmodulin protein kinase, protein kinases A and C, etc.   Among the responses of these kinases are phosphorylation/ dephosphorylation of cytoskeletal elements such as microtubule associated proteins ("MAPs") and intermediate (neuro) filaments which can reconfigure intra-neuronal structural architecture. In Paramecium, adaptive responses of these membrane-second messenger-cytoskeletal signaling cascades involve complex activities of cilia; in neurons they include cognitive functions such as learning and memory which apparently involve regulation of synaptic strengths and architectures. The cytoskeleton is at least circumstantially linked to cognitive function by a line of evidence which includes the following.

Mileusnic et al (1980) correlated production of MT subunit protein ("tubulin") and MT activities with peak learning, memory and experience in baby chick brains. Cronley-Dillon et al (1974) showed that when baby rats begin their critical learning phase for the visual system (when they first open their eyes), neurons in the visual cortex begin producing vast quantities of tubulin. Tubulin production is drastically reduced when the critical learning phase is over (when the rats are 35 days old). Moshkov et al (1992) showed structural and chemical changes in goldfish brain neuronal cytoskeleton following sensory stimulation. In gerbils exposed to cerebral ischemia (lack of brain oxygen supply), Kudo et al (1990) correlated the amount of reduction in dendritic MAP-2 with the degree of cognitive impairment. Bensimon and Chernat (1991) found that selective destruction of brain MT by the drug colchicine caused cognitive defects in learning and memory which mimic the clinical symptoms of Alzheimer's disease, in which the cytoskeleton becomes entangled. Geerts et al (1992) showed that sabeluzole, a memory-enhancing drug, increases fast axoplasmic transport. Matsuyama and Jarvik (1989) have proposed that Alzheimer's is a disease of MT and MAPs. In specific hippocampal regions of the brains of schizophrenic patients, Arnold et al (1991) found distorted neuronal architecture due to a lack of 2 MAPs (MAP-2 and MAP-5).

Cognitive mechanisms are often approached through the study of long-term potentiation ("LTP"), models of learning in mammalian NMDA (n-methyl-d-aspartate) hippocampal neurons which involve pre- and post-synaptic enhancement of synaptic function. Post-synaptic learning mechanisms involve second messenger-kinase activation, which includes cytoskeletal regulation. For example, Silva et al (1992a; 1992b) have shown calcium-calmodulin protein kinase to be essential for LTP. Aszodi et al (1991) have demonstrated protein kinase A to be involved in learning. Halpain and Greengard (1990) have shown that post-synaptic activation of NMDA receptors induces rapid dephosphorylation of dendrite-specific MAP2. Aoki and Siekevitz (1988) linked learning to protein kinase C mediated MAP2 phosphorylation/dephosphorylation.   Bigot and Hunt (1990) showed that NMDA and other excitatory amino acid neurotransmitters caused redistribution of intraneuronal MAPs and Rasenick et al (1990) and Wang and Rasenick (1991) have shown that G proteins directly link with microtubule proteins. Surridge and Burns (1992) demonstrated that the lipid second messenger phosphatidylinositol binds specifically to MAP2, a coupling which may "constitute a link by which extracellular factors influence the microtubule cytoskeleton." Kwak and Matus (1988) showed that denervation caused long-lasting changes in MAP distribution, and Desmond and Levy (1988) showed that LTP involved cytoskeletal-mediated shape changes in dendritic spines. Aoki and Siekowitz [1988] have studied dendritic spine synaptic plasticity in development and find that synaptic down-regulation occurs in conjunction with depolymerization of microtubules in the main dendrite. They find also that signaling and regulation for dendrite spine synapse function depends on phosphorylation of dendrite-specific MAP2. Theurkauf and Vallee [1983] have found MAP2 to be "the major substrate for endogenous cyclic AMP-dependent phosphorylation in cytosolic brain tissue", and concluded that MAP2 phosphorylation "may be an important reaction in response to neurotransmitter stimulation." Thus, numerous mechanisms exist for coupling membrane synaptic events and cytoskeletal activities. Lynch and Baudry (1987) have proposed that proteolytic digestion of the sub-synaptic cytoskeleton, followed by its structural reorganization, correlate with learning and Friedrich (1990) has formalized a learning model of synaptic cytoskeletal restructuring. Synaptic regulation, both in learning and steady state, also depends on material (enzymes, receptors, neurotransmitters, etc.) synthesized in the cell body and transported along axons and dendrites by contractile MAPs attached to MT. Similar mechanisms involving actin, synapsin, and other cytoskeletal polymers are involved in neurotransmitter release. Figures 6-6 and 6-7 demonstrate cytoskeletal structures in relation to synapses and dendritic spines.

Some documented cytoskeletal functions do, by themselves, exhibit adaptive behaviors relying on molecular logic. For example, in "dynamic instability", MT can alternate erratically between growing and rapidly shrinking phases. This mechanism ("a generator of diversity") is utilized for probing and reorientation within cells ranging from growing neurons to motile amoeboid movement in macrophages. Such adaptive behavior (in which a stabilizing factor may be considered a positive attraction stimulus) can be observed *in vitro* without membrane or genetic input and can also be computer simulated (Hotani et al, 1992).

F:6-6

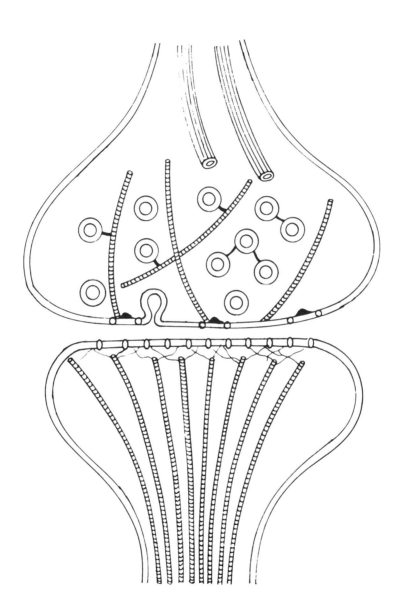

Schematic of cytoskeletal structures in synapse. Top: Pre-synaptic axon with MT, synapsin, neurotransmitter vesicles. Bottom: Post-synaptic dendrite with receptors, linking proteins, filaments/MT.

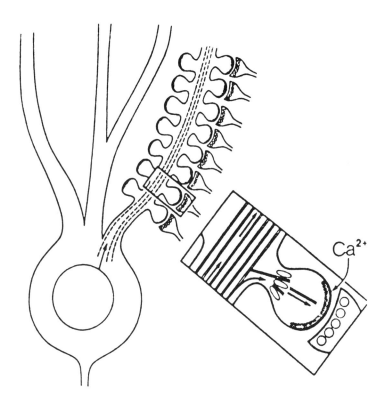

Schematic of neuronal dendrites and dendritic spines with internal cytoskeleton in one dendrite. Rectangle: Closeuup of dendrite and dendritic spine with cytoskeleton visible.

Maintenance and modulation of synaptic function also depend on axoplasmic transport, an MT based conveyor system which supplies structural components to synapses (Lasek, 1981; Ochs, 1982). Contractile motor MAPs (dynein, kinesin) attached at specific sites on microtubules pass materials in a cooperative bucket brigade which achieves a velocity of 400 mm/day (Figure 6-8). The contractile sidearms hydrolyze ATP for energy, but the mechanism of orchestration and signaling is unknown. Material can travel in opposite directions along a single microtubule, and the same motor MAPs can change directions under different conditions. Other organized activities in cytoskeletal structures help control the molecular machinery of cell division, growth, differentiation, formation of synapses and dendritic spines.

Thus varied adaptive behaviors in cells ranging from paramecia to mammalian cortical neurons (reversal of ciliary beating, altered synaptic structure and efficacy, induction of genetic responses, etc.) share a common feature. In each the cytoskeleton provides the "missing link," a communication network among membrane, second messenger and nuclear genetic elements (in neurons this may traverse a rather large distance). The role of the cytoskeleton, however, is generally held to be passive and subservient. Cytoskeletal components are considered to constitute merely the cell's structural scaffolding and mechanical conveyor belt system controlled by membrane regulated ion fluxes, second messengers, and genetic influences. Despite this prejudice, there are reasons to believe that cytoskeletal activities may underlie intelligence and cognition in cells ranging from lowly paramecia to human cortical neurons. We will discuss those reasons, and describe mechanisms by which cytoskeletal microtubules could conduct and process information.

6.2   Cytoskeletal cognition

6.2.1   The quasicrystalline structure of living matter

The concept of cytoskeletal computation is based on the observation that cytoplasmic interiors of living cells, particularly nerve cells, are not watery soups. The current picture of cell interiors is one of a highly organized network in which cell water is bound and governed by cytoskeletal structures rather than a freely diffusing solution of macromolecules. Two lines of evidence support this concept. The first is the cytoskeleton which

F:6-8

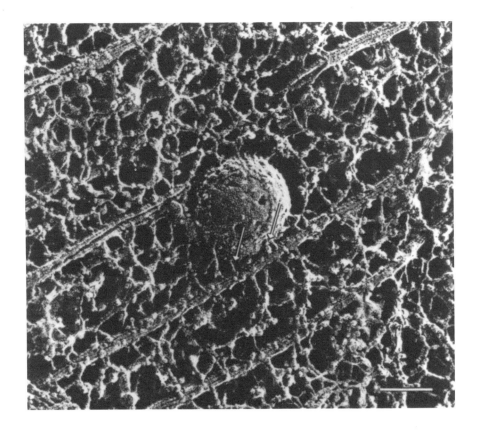

Axoplasmic transport: vesicle transported through axoplasm by contractile activities of dynein and kinesin MAPs (arrows) attached to MT. Such transport maintains and regulates synapses. Scale bar: 100 nm.

interpenetrates the cytoplasm and defines its architecture and function (Figure 6-9). Some cytoskeletal components are in polymerization equilibrium, interconverting cytoplasm between "sol" (liquid, solution) and "gel" (gelatinous, viscous) states. This "sol/gel" equilibrium (mediated by Ca^{2+} and Mg^{2+} effects on cytoskeletal actin and other proteins) indicate that interiors of living cells operate very close to the solid-liquid phase transition as a liquid quasicrystal. Theoretical findings predict optimal computational capabilities for distributed dynamical systems near this phase transition ("the edge of chaos," Langton, 1990).

Another line of evidence supporting the possibility of a computational cytoplasm includes data from nuclear magnetic resonance (NMR), neutron diffraction and other techniques which show a high degree of bound water in the cytoplasm. Water molecules abutting cytoskeletal (and other) cytoplasmic surfaces are attracted to these surfaces and are restricted in their motions. As a result, a greater proportion have four (rather than three or less) hydrogen bonds with their neighbors. Such water is called "vicinal" water, and has unusual properties compared to "bulk" water: lower density, greater heat capacity and greater viscosity. Further, such vicinal water may cooperatively oscillate with coherent excitations in cytoskeletal proteins.

Further circumstantial evidence for cytoskeletal computation and communication stems from the spatial distribution of discrete sites/states in the cytoskeleton. For example, subunits in microtubules have a "density" of about 10^{17} per cm^3, very close to the theoretical limit for charge separation (Gutmann, 1986). Thus, cytoskeletal arrays have maximal density for information storage via charge, and the capacity for dynamically coupling that information to mechanical and chemical events via conformational states of proteins (i.e. Fröhlich's mechanism of dipole excitations coupled to conformational state).

6.2.2 Connectionism

The use of connectionist models of functional brain organization has advanced understanding of cognition and linked neuroscience to computer science. "Neural networks", which approximate cortical neurons as parallel processors with variable lateral interconnections, may be simulated on computers and can provide a working model of some aspects of brain function (e.g. Hopfield, 1982). A key concept relating neural network dynamics to neuroscience was introduced by Hebb (1949). He hypothesized that spatially organized neural networks called "cell assemblies" function as reverberatory

F:6-9

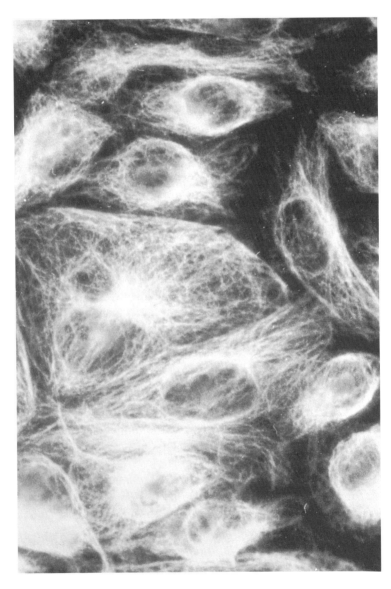

Immunofluorescence of tissue culture cells stained with tubulin antibody. White lacy structures are microtubules. **Courtesy of Larry Vernetti.**

circuits which constituted elements of thought.   Gazzaniga's "modules," Minsky's "agents", Freeman's "cartels" and other conceptualized functional entities are examples of more recent, comparable proposals for anatomic neural networks (Rasmussen et al, 1990). In Hebb's view, an individual neuron could participate in many cell assemblies just as an individual person could be a member of various social groups. Using motor tasks in monkeys, Montgomery et al (1992) showed that individual neurons do participate in multiple networks, rapidly switching among them ("the multipotential neuron"). By strengthening or reinforcing repeatedly used connections ("synaptic plasticity"), Hebb suggested that recognition, learning and problem solving occurred through lowered thresholds of specific loops. By assigning energy levels to threshold loop patterns ("landscapes"), mathematical solutions could be applied to neural net configurations (Hopfield, 1982).

There are, however, several inconsistencies in the analogy between neural nets and brain function. The first is that neurons and their synapses are extremely complex and by themselves resemble computers more than fundamental switches. Their dynamic complexity suggests that neurons utilize "lower" layers which may include dendritic processing, branch point conductance and molecular dynamics (Scott, 1977). A second inconsistency in the brain/artificial neural network analogy is that some important neural nets require "back-propagation," tuning of internal parameters from output conditions. A third is the question of how a "multipotential" neuron can rapidly (? simultaneously)switch among multiple networks in which it participates. All of these inconsistencies may be resolved by consideration of the cytoskeleton, intracellular parallel networks of protein polymers which regulate synapses and perform other important functions.

In the cytoskeleton, filamentous bridges among parallel MTs and /or neurofilaments could serve functions comparable to the recognition automata of Reeke and Edelman (1984). To take maximal advantage of parallelism, these authors have modelled two parallel automata which communicate laterally and have distinct and complementary personalities. One model automaton ("Darwin") is highly analytical, keyed to recognizing edges, dimensions, orientation, color, intensity, etc. The other ("Wallace") is more "gestalt" and attempts to merely categorize objects into preconceived classifications. Lateral communications between the two parallel automata resolve conflicting output and form an associative memory. In the cytoskeleton, filamentous bridges among parallel MTs and/or neurofilaments could serve comparable functions.

Protein as Finite State Machine

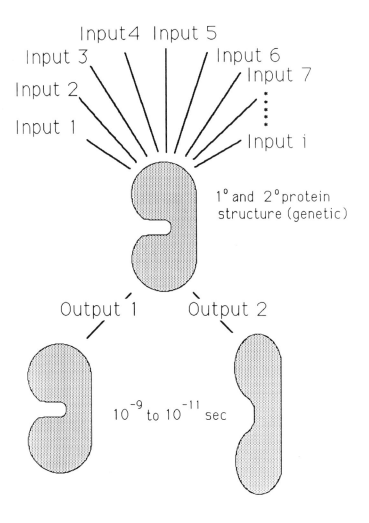

6.2.3 Cognitive Hierarchy--A Cytoskeletal Basement Dimension

Multi-level nets are consistent with connectionist brain models in which neural net "assemblies", "modules", etc. fit in an overall cognitive hierarchy of parallel layers of brain/mind organization (Churchland and Sejnowski, 1988). The highest cognitive level in such models is global brain function which correlates with awareness, thought, or "consciousness" (although some would argue for even higher-level social consciousness among people, political parties, societies, etc.). The second level appears to be comprised of anatomically and functionally recognizable brain systems and centers (i.e. respiratory center, satiety center, etc. [Somjen, 1983]). At intermediate levels, maps such as the motor and sensory homunculi represent the body and outside world. Finally, the lowest level is thought to be the neural synaptic network, module or cartel which may operate cooperatively by utilizing dense interconnectedness, parallelism, associative memory and learning due to synaptic plasticity.

The lowest level of brain/mind organizational hierarchy, and potential correlate of the binary switch in computers, is generally considered to be the neuronal synapse. However, we suggest that information processing in the cytoskeleton could provide a basement level. By subserving synaptic connectionism, such information processing can provide a lower level in a hierarchy of cognitive processes from whose highest level emerges consciousness. Further, retrograde patterns travelling through the cytoskeleton could serve a role analogous to back-propagation in some two-dimensional or three-dimensional artificial neural networks. In Section 6.3.3, we describe 2-dimensional simulation of such retrograde patterns travelling through microtubules.

6.3   Cytoskeletal Computation - Current Theoretical Model

Information representation and processing in cytoskeletal lattices depend upon representation of information quanta ("bits") in lattice subunits (i.e. tubulin in MT). Protein conformational states offer the capabilities for dynamic and interactive representation of information units. Figure 6-10 conceptualizes protein as finite state machine.

6.3.1 Protein Conformational Dynamics

In their natural state, proteins are dynamic structures which undergo conformational motions over a wide range of time and energy scales. Conformational changes related to protein function occur in the nanosecond (10^{-9} sec) to 10 picosecond (10^{-11} sec) time scale. Related to cooperative movements of protein sub-regions and charge redistributions, these changes are linked to protein function (signal transduction, ion channel opening, enzyme action etc.) and may be triggered by factors including phosphorylation, ATP or GTP hydrolysis, ion fluxes, electric fields, ligand binding, and neighboring protein conformational changes. In the case of tubulin within MT, we propose that such programmable and adaptable states can represent and propagate information. Signaling in MT-tubulin was shown experimentally by Vassilev et al (1985), and theoretical approaches which can describe such propagation include elastic self-trapping modes such as solitons, coherent phonons, and polarization waves. Solitons are quantized packets of energy (and information) which travel with minimal dissipation in water (canal waves, ocean tidal waves), optical fibers (carrying information such as telephone data), and other media. Conformational solitons have been predicted to occur in biological materials such as actin-myosin, DNA, MT, and MT-clathrin. These theoretical cases predict that energy is supplied by dephosphorylation or hydrolysis of ATP or GTP, and that the energy/information packet is utilized in muscle contraction, DNA replication, and MT signaling and polymerization, respectively.

Fröhlich (e.g. 1975), proposed that protein conformational changes are coupled to charge redistributions such as dipole oscillations within specific hydrophobic protein regions. (Such protein hydrophobic regions are also where general anesthetic gas molecules are thought to act by preventing protein conformational responsiveness.) Fröhlich further proposed that a set of proteins connected in an electromagnetic field such as within a polarized membrane (or polar polymer like a microtubule) would be excited coherently if biochemical energy such as protein phosphorylation or ATP or GTP hydrolysis were supplied. Coherent excitation frequencies on the order of 10^9 to 10^{11} Hz (the time domain for functional protein conformational changes, and in the microwave or gigaHz spectral region) were deduced by Fröhlich who termed them acousto-conformational transitions, or coherent phonons. Other aspects of Fröhlich's model include metastable states (longer-lived conformational state patterns stabilized by local factors) and polarization waves (traveling regions of dipole coupled conformations).

Experimental evidence for such coherent excitations includes observation of gigaHz-range phonons in proteins, sharp-resonant non-thermal effects of microwave irradiation on living cells, gigaHz induced activation of MT functions in rat brain, and long-range regularities in cytoskeletal structures, such as the super-lattice attachment pattern of MAPs on MT. Samsonovich et al (1992) showed that experimentally observed patterns of MAP attachment sites on MT can be simulated and possibly derive from self-localized coherent phonon excitations. Coherent conformational excitations may provide a clocking mechanism for MT information processing in the context of cellular automata.

6.3.2 Cellular Automata

Computation involves interactive signals or patterns in a lattice structure. Von Neumann described such systems (which include computers as special cases) as "cellular automata," which consist of a large number of identical "cells" connected in a uniform pattern. The essential features of cellular automata (of which Conway's "Game of Life" is one popular example) are: 1) at a given time, each cell is in one of a number of finite states (usually two for simplicity); 2) The cells are organized according to a fixed geometry; 3) Each cell communicates only with other cells in its neighborhood--the size and shape of the neighborhood are the same for all cells; and 4) There is a universal clock. Each cell may change to a new state at each tick of the clock (or "generation") depending on its present state and those of its neighbors. The neighbor "transition" rules for changing states, though simple, can lead to complex, dynamic patterns manifesting chaos, fractal dimensions, partial differential equations, and computation. Patterns which move through the lattice unchanged are called "gliders"; Von Neumann proved mathematically that gliders travelling through a sufficiently large cellular automaton can solve virtually any problem.

Cellular automata are theoretically advantageous for molecular computing because the internal connections are intrinsic to the material, external connections need only occur in one limited region, and computation can occur by local interactions with speed dependent on the clocking frequency. Conrad (1973) used the concept of "molecular" automata within neurons as an information processing system subserving the brain's synaptic connectionism.

We have applied cellular ("molecular") automata principles (using Fröhlich's coherent 10^{-9} to 10^{-11} sec excitations as a clocking frequency) to the dynamic conformational states of tubulin within cytoskeletal microtubules. We hope to understand and explain real-time control, self-organization, communication and computation in living cells, and to propose precursors to molecule computing paradigms.

6.4 Molecular Automata in Microtubules

6.4.1 Single Microtubule Automata

"Cellular" (molecular) automata (as represented in 2 dimensions) require a lattice whose subunits can exist in two or more states at discrete time steps, and transition rules which determine those states among lattice neighbor subunits. Tubulin conformations within MT lattices can provide such states, and neighbor-tubulin interactions (represented by dipole coupling forces) may provide appropriate transition rules. MT automata were described in Hameroff et al (1989) and Rasmussen et al (1990). A rough estimate for the time steps, assuming one coherent "sound" wave across the MT diameter (D ~ 25 nm) and $V_{sound}=10^{5}$ cm/s, yields a clocking frequency of approximately 3 x 10^{10} Hz, and a time step of 2.5 x 10^{-11} sec. Thus Fröhlich's coherent excitations can provide a "clocking frequency" for MT conformational automata.

Conformation of individual tubulin subunits at any given time depends on "programming" factors including initial conformational state, primary genetic structure, binding of water, ions, or MAPs, bridges to other MT, post-translational modifications, phosphorylation state, and mechanical and electrostatic dipole forces among neighboring subunits. We consider only electrostatic dipole forces among neighboring tubulin subunits. Figure 6-11 shows MT structure and a 7-member MT automata neighborhood: a central dimer surrounded by a tilted hexagon of 6 neighbor dimers. The two monomers of each dimer share a mobile electron which is oriented either more toward the α-monomer ("alpha state") or more toward the ß-monomer ("beta state") with associated changes in dimer conformation at each time step.

F:6-11 and 6-12

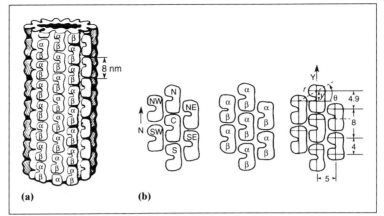

(a) MT Structure from x-ray crystallography (Amos and Klug, 1974). Tubulin subunits are 8 nm dimers comprised of α and β monomers. (b) MT automata neighborhood: left, definition of neighborhood dimers; center, α and β monomers within each dimer; right, distances in nm and orientation among lattice neighbors.

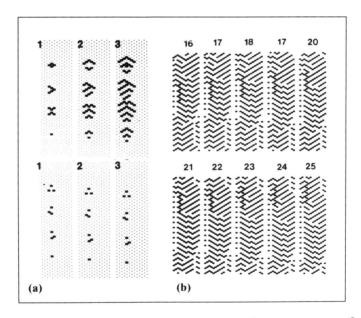

MT automata models with MT displayed as rectangular grids (Rasmussen et al, 1990). Black elements are β conformational state tubulin dimers; white elements are α states. (a) Top: Three successive time steps for four objects (i. "dot glider", ii. "spider glider", iii. "triangle glider", iv. "diamond blinker") at threshold ± 1.0, moving downward, leaving traveling wave patterns. Below: Three time steps for a "dot" and three other gliders at higher threshold ± 9.0. The gliders travel downward without a wake.

The net electrostatic force from the six surrounding neighbors acting on a central dimer can then be calculated as:

$$f_{net} = \frac{e^2}{4\pi\epsilon} \sum_{i=1}^{6} \frac{Y_i}{r_i^3} \qquad (6\text{-}1)$$

where y_i and r_i are defined as illustrated in Figure 6-11, e is the electron charge, and ϵ is the average protein permittivity. Neighbor electrostatic dipole coupling forces may be found in Rasmussen et al (1990).

To simulate MT automata, cylindrical MT structure is displayed as 2-dimensional rectangles whose edges (adjacent protofilaments) are contiguous. To avoid boundary conditions, end borders also communicate so that a torus is modeled. MT subunit dimer loci are in either α state (white) or β state (black). At each "nanosecond" time step, forces exerted by 6 surrounding neighbor subunits are calculated for each dimer. If the net force exceeds a threshold, a transition ($\alpha \rightarrow \beta$, $\beta \rightarrow \alpha$) occurs. For example, a threshold of ± 9.0 means that net neighbor forces greater than $+9.0 \times 2.3 \times 10^{-14}$ Newtons will induce an alpha state, and negative forces of less than $-9.0 \times 2.3 \times 10^{-14}$ Newtons will induce a β state. Threshold may represent temperature, pH, voltage gradients, ionic concentration, genetically determined variability in individual dimers, binding of molecules including MAPs and/or drugs to dimer subunits, etc. Effects of varying thresholds on MT automata behavior are shown in Figure 6-12a. Behaviors include gliders, traveling and standing wave patterns, oscillators, linearly growing patterns, and frozen patterns (perhaps suitable for memory). Asymmetrical thresholds ($\alpha \rightarrow \beta \neq \beta \rightarrow \alpha$) result in bidirectional gliders (Figure 6-12b).

In the next sections, we consider network properties of MT automata interconnected by MAPs.

6.4.2 MT-MAP Network Formation

Architecture and functions of cells depend heavily on cytoskeletal lattices, consisting largely of MT-MAP networks.   MAPs stabilize MT, preventing disassembly and promoting assembly.   We investigate MT stabilization by MAP binding on adjacent, parallel MT as an initial step in construction of a MT-MAP network.

We assume that MAP attachments on MT assume periodic distributions and MAP binding to a specific tubulin increases its affinity to its six surrounding neighbors; this neighborhood then becomes resistant to disassembly (Hotani et al, 1992).   Maintaining a periodic distribution of MAPs, only a finite number of MAP attachment sites are available for each region on a MT. The number of MAPs per region and their particular location(s) within that region may be related to the reduction of the free energy value which, in turn, may be related to the stabilization of MT sub-assemblies. Using a "mobile finite automata" technique, we have shown a number of MAP distributions and their calculated free energy values (Figure 6-13; Hameroff et al, 1992). The efficiency in free energy reduction depends on and determines both MAP location and number.

6.4.3 Learning Via Optimization of MAP Connection Sites

We next consider adaptive behavior in simple MT-MAP automata networks demonstrating input/output learning found in artificial neural networks ("ANNs").   Two MT automata are interconnected via MAP connections capable of transmission of signals (sequences of α or β conformational states) from one MT to another.   MAP connection sites vary randomly in an evolutionary optimization process, and two inputs and one output are defined as regions on the two MTs.   For two different sets of MT1 and MT2 input pairs (automata patterns), a desired or correct output pattern is predefined. After each sequence of time steps (sufficient for patterns to propagate from the input areas to the output area, with transmission across MAP connections) the output area pattern is compared to the desired output pattern using a mathematical formulation of error called the "Hamming distance" ("H_d": the number of digit positions in which two binary words of the same length differ). Allowing random MAP connection topologies between the two MT, the most efficient topology (that which yields the lowest Hamming distance) is selected as the "mother system" after each time step sequence.   At the next step other "daughter" topologies are randomly created.   When one daughter performs better than the original system (lower Hamming distance), this connection topology becomes the mother system for the next generation (Rasmussen et al,

F:6-13 and 6-14

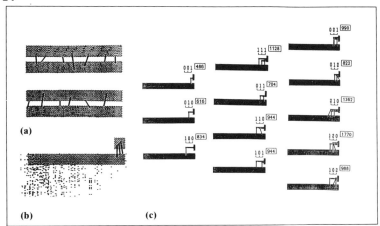

Simulation of MAP connections and MT network initiation. (a) Some possible MAP connection sites and connection topologies, (b) MT disassembly until stabilized by MAPs, (c) MT MAP quantity and connection topologies which promote neighbor MT assembly and stability.   Three possible "rows" of MAP connection sites are considered, negative free energy values were calculated and shown in rectangles.

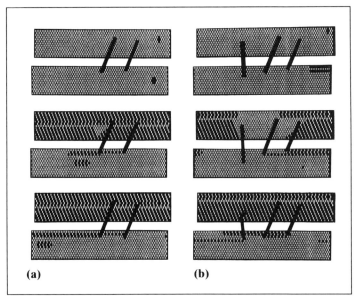

(a) MTA net learning process. The connection topology with MAPs at dimer locations MT_1: (60,6) → MT_2: (55,2) and MT_1 (47,4) → MT_2 (41.3) satisfies an input output map (not shown) with $H_d = 0$. Dynamics shown at time steps 0, 43, and 66. (b) Same MT net at later stage of learning process after it has adapted to yet another input/output map.

F:6-15 and 6-16

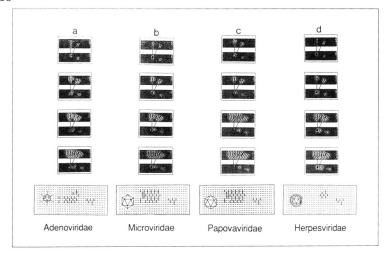

Temporal evolution of MT automata patterns at time 2, 4, 6 and 8 with MT2 automata patterns for 4 viruses studied. The bottom row (row 5) is an enlargement of an MT2 pattern area. (a) MAP-left closed, MAP right open; (b) MAP left open, MAP right closed; (c) both MAPs closed; (d) both MAPs open.

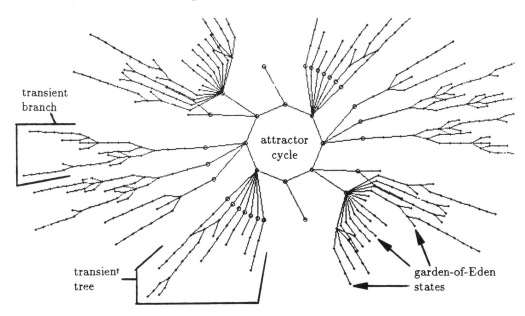

Basin of attraction for a cellular automaton, representing all possible behaviors leading into the attractor. Each automaton lattice configuration corresponds to a state and corresponds to a node in the graph. The edges in the graph indicate the possible state transitions towards the attractor cycle.

1990).

Input-output maps $(I_{11}, I_{21}) \rightarrow O_{21}$ and $(I_{12}, I_{22}) \rightarrow O_{22}$ were used for learning trials. The connection topology $C_1 \rightarrow C_2$ which satisfies the first input-output map is shown in Figure 6-14a together with the network patterns at time 0, 43, and 66 ($H_d = 0$). In Figure 6-14b the connection topology $C_k \rightarrow C_4$ which satisfies the second input-output map is shown together with corresponding MT patterns at time 0, 23, and 66 ($H_d = 0$). This topology also satisfies the first input-output map. Because of perturbation-stable glider patterns, the MT network is also able to "associate" patterns, that is produce correct outputs from inputs which are similar, but not identical, to the original output-coupled inputs.

6.4.4 MT-MAP Network Recognition Via Membrane Coupled MAP Regulation

This model system considers MAPs attached at fixed loci between 2 MT automata as switches ("open" or "closed"). Each MAP's state ("open" or "closed") is determined by membrane receptor activation (i.e. by binding of neurotransmitter molecules to post-synaptic neuronal receptors) coupled to the MAPs by second messengers (e.g. G-proteins, [Ca^{++}], cyclic AMP, protein kinases, MAP phosphorylation, etc.). Lahoz-Beltra (Hameroff et al, 1992) used 3 binary receptor inputs, modeling the binding (1) or not (0) of neurotransmitter molecules. Receptors regulate MAP1 and MAP2 which each interconnect MT1 and MT2. If either MT1-MAP connection tubulin is in α state and the MAP is open, the corresponding MT2-MAP connection site will be perturbed (if α, then ß; if ß, then α). MAPs have a critical phosphorylation threshold, above which MAPs are "open." Outputs occur on MT2 distal to both MAP attachments.

This model network was used in a training and recognition exercise. As an arbitrary example, characteristics of virus structure were expressed in 3 binary categories as inputs to the receptors. Four types of viruses: microvirus, papovavirus, adenovirus, and herpes virus were categorized by these 3 characteristics [naked = (1), enveloped = (0); number of capsomer subunits in protein capsids: $<32=(1)$, $>32=(0)$; diameter of virus (nm): $<50=(1)$, $>50=(0)$]. Training sets and MT2 automata outputs for the four virus patterns are presented and results of a virus recognition test are shown in Figure 6-15; five of six viruses were correctly identified.

6.4.5 Artificial Neural Network Models Utilizing Cytoskeletal Signaling

Artificial neural networks (ANNs) are computer designs for learning and computation based on processing units that are inspired by biological neurons. Although some ANN activities seem biologically plausible, those that require backwards feedback along each forwards connection have not appeared biologically plausible without retrograde internal signals within neurons (i.e. axon to dendrite). ANN paradigms that utilize backwards feedback include back-error propagation, sigma-pi and RCE architectures, and adaptive resonance theory. Backwards feedback across synapses in biological neurons has now been shown likely to occur via nitric oxide. Backwards neuronal feedback signals, if provided by membrane potentials, would require each forwards connection to have a complementary backwards neuronal synaptic connection. If provided by retrograde axoplasmic transport, backwards feedback signals would be rather slow.

As described by Dayhoff (Hameroff et al, 1992), the cytoskeleton may be capable of carrying fast conformational feedback signals backwards through a network of neurons. Axonal branches convey error signals $d(R_{target}-R_{output})$ back to the axon hillock where they are averaged and transmitted to dendritic synapses to adjust synaptic strengths (s) accordingly. The strength of the backwards signal (e.g. automata gliders) could be in the number of gliders or in the size of the MT bundle that carries that signal. Bundles might grow, shrink and change stability (via pattern induced MAP binding, nucleation/stabilization of adjacent MTs, post-translational modifications, etc.) to modulate feedback and mediate learning.

6.4.6 Functional Capacity of Microtubule Automata

Assuming MT conformational automata gliders and patterns exist, what functions could they have? They could represent information being signaled through the cell. Glider numerical quantities and patterns may manifest signals, binding sites for ligands, MAPs or material to be transported. Frozen patterns may store information in a memory context, information may become "hardened" in MT by post-translational modifications, and/or MT automata patterns could transfer and retrieve stored information to and from neurofilaments via MAPs.

MT automata gliders travel one dimer length (8 nm) per time step (10^9

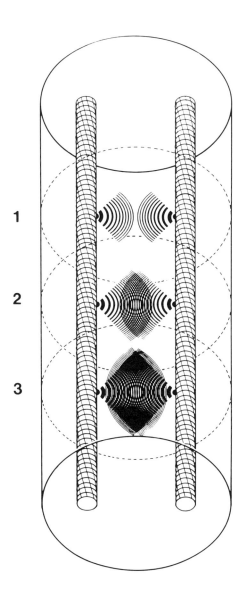

Coherent excitation of MT causing calcium ion waves, coherent sol-gel in cytoplasmic density patterns which can interfere in a holographic mechanism.

to 10^{-11} sec) for a velocity range of 8 to 800 meters per sec., consistent with propagating solitons or phonons which travel within neurons concomitantly with membrane potentials. Thus traveling MT automata patterns or gliders (equivalent to solitons, phonons or Fröhlich depolarization waves) may propagate in the cytoskeleton in concert with membrane depolarizations and ion fluxes. Long range cooperativity, resonances and phase transitions among spatially arrayed MT automata patterns may occur, and membrane related voltages, ion fluxes or direct links could induce transient waves of conformational switching or lowered threshold along parallel arrayed MT. Consequently, activity of a particular cell could directly elaborate patterns within that cell's MT automata, phenomena important in emergent cognitive and behavioral functions ranging from simple organisms to human brains.

This coupling can lead to a view of the brain/mind as a hierarchy of nested automata with a previously overlooked basal dimension. For example, artificial intelligence/roboticist Hans Moravec (1987) has calculated the computing power of the brain based on the classical "neural net" assumption that each neuron-neuron synaptic connection is a fundamental binary switch. He assumes 40 billion neurons which can change state hundreds of times per second, resulting in roughly 4×10^{12} "bits" per second. However, rather than a simple switch, each neuronal synapse is extremely complex and dynamically controlled and modulated by the cytoskeleton. (Further, single-cell organisms like amoeba and paramecium perform complex tasks without benefit of synapse, brain, or nervous system.) Considering the cytoskeleton as a sub-dimension in a hierarchy of nested automata increases the calculated information capacity of the brain/mind immensely. Assuming parallel MT spaced about 100 nanometers apart and the volume of brain which is neuronal cytoplasm to be about 40 percent leads to approximately 10^{14} MT subunits in a human brain (ignoring neurofilaments and other cytoskeletal structures). If each subunit can change state every 10^{-9} to 10^{-11} seconds (as per Fröhlich's theory), the brain's MT would have an information capacity of roughly 10^{23} to 10^{25} bits per second! This would allow for massive parallelism and redundancy as interaction conformational state patterns compute at a sub-level of the brain's hierarchy of parallel information processing systems. The branching patterns of neural axons and dendrites have been likened to trees. MT and related filaments may represent a cytoskeletal forest within those trees.

6.5   Logics

6.5.1   Logic Outline

Current approaches in neurocomputing are based on Boolean and/or fuzzy logic. The McCullough-Pitts binary neuron concept was proposed in 1943 as a biological model of the neuron and has been used to built networks for computing Boolean logical functions (McCulloch and Pitts, 1943). Many variations of "neuronal" processing elements based on this concept have been proposed (Hecht-Nielsen, 1990). Lahoz-Beltra et al (1993) have formulated a model of Boolean molecular logic in MT-MAP networks. However, scientific results in physics, particularly the Heisenberg uncertainty principle, open a new view in logic, and have led to the development of multi-valued logics. For example, a strong scientific approach to fuzzy logic as a multi-valued logic was developed by Zadeh in 1965 (Zadeh, 1965). After further research by many scientists, there is currently strong scientific and engineering implementation of fuzzy logic and fuzzy systems in neural networks (Kosko, 1992). Using ideas from biology, Koruga describes an "infinite value" aleph ($\aleph_0$) logic which encompasses both Boolean and fuzzy logic (Koruga & Hameroff, 1992).

6.5.2 Bio-Mathematical Fundamentals

There are three main periods for a human being. The first period is the uniting of a male gamete or sperm with a female gamete or ovum to form a single cell called a zygote. The second period is embryological development, and the third period begins after birth when a human being interacts with the environment. Embryology is important because it provides a knowledge of the relationships of body (brain) structures as intelligent hardware.

Although human development and genetics are still incompletely understood, developmental processes depend upon a precisely coordinated interaction of nuclear genetic, cytoplasmic and environmental factors.   The nervous system develops from a thickening called the neural plate which appears during the third week of embryological development. The neural plate engenders the neural tube, the upper part of which develops into the brain.

During the fourth week, three main brain vesicles are formed: the forebrain, the midbrain, and the hindbrain. During the fifth week, the forebrain divides into two vesicles, and at the end of embryology there are five secondary brain vesicles, all of which become various brain components.

Five main facts are important: (1) the embryonic period is based on "exact" spatio-temporal relationships (high level coordination in spacetime); (2) each differentiated cell (neuron) has the same genetic material as the initial cell (zygote); (3) although the genetic materials are the same, cytoplasmic materials differ from the zygote cell and also differ from cell to cell within an organism; (4) the organism and brain have left and right symmetry; and (5) there is an inversion in information processes (left side of brain controls the right side of body and vice versa).

In this case we can write:

$$C_{zl} = \{C_{nz}, C_{cz}\} = 1 \qquad (6\text{-}2)$$

$$C_{nz} = C_{nl}^{1,2} + C_{n2}^{2,3} \ldots C_{nm}^{k,l} \qquad (6\text{-}2a)$$

$$C_{nz} = C_{nl}^{1,2} = C_{n2}^{2,3} \ldots = C_{nm}^{k,l} \qquad (6\text{-}2b)$$

$$C_{cz} = C_{cl} + C_{c2} + \ldots + C_{cm} = \int_{1}^{m} f[x(3),t]dxdt = 1 \qquad (6\text{-}3)$$

$$O_{w} = L_{s} + R_{s} = 1 \qquad (6\text{-}4)$$

Where

C$_z$:   Represents zygote as an information entity
C$_{nz}$:   Information of zygote's nucleus
C$_{cz}$:   Information of zygote's cytoplasm
C$_{nl}^{km}$:   Information of cell's nucleus in "m" interactions of differentiation, in "k" vesicles, in "l" cells
C$_{cj}$:   Information of cell's cytoplasm in "m" interactions of differentiation, in "k" vesicles, in "j" cells
x(3):   Three dimensional space [x$_1$, x$_2$, x$_3$]
t: time
O$_w$:   Represents organism as information entity of whole
L$_s$:   Left side of body (brain)
R$_s$:   Right side of body (brain)

From a mathematical point of view, it is interesting how is it possible to satisfy Equations 6-2a, 6-2b and 6-3. Equations 2a and 2b could be satisfied only in two cases. First if C$_{nz}$=0, C$_{n1}$=0,...C$_{nm}$=0 which doesn't make sense. Second if C$_{nz}$= $\aleph_0$, C$_{n1}$= $\aleph_0$, C$_{n2}$= $\aleph_0$ . . . C$_{nm}$,= $\aleph_0$ because (Dauben, 1979):

$$\aleph_0 + \aleph_0 + \aleph_0 + \ldots + \aleph_0 = \aleph_0 \qquad (6\text{-}5)$$

For interpretation of Equations 6-3 and 6-4, the key point is the mechanism of cell division. The fundamental question is the function of the centrosome in this process. Experimentally it was shown that without the centriole, as a main component of the centrosome, assembly cannot proceed "*de novo*" in mammalian somatic cell (Bailly and Borners, 1992). Interpretation of experiments on this subject only could be that the centrosome cycle might provide the cell with same kind of "logic" capable to monitor time and mass increase and coupling the cell growth and the cell division. Embryological process (E) is a massively and highly parallel dynamical system with discrete steps. If we use a discrete dynamical system in which time evolution is described by discrete steps it is possible to find a function E which maps a space C (Cell) unto itself and the iterates of E that is E^1, E^2 ...E^n, where E^{n+1} (c)=E(E^n(c)). The invariant set as a subset C$_c$ (cytoplasm) of C such that E(C$_c$) $\subset$ C$_c$ we can classify according to how the points near the invariant set behave under E. In this case the invariant set C$_c$ we call the "*cell-attractor*" if there is a region O$_c$ (organizing center) such that if c $\in$ O$_w$, then E^n (c) gets closer and closer to O$_w$ as "n" increases. We want a measure such that H$_k$(C-Cc)=0 and $H_k(E^{-1}(C_s))=H_k(C_s)$ for each measurable subset C_s of C_c (C_s-centrosome). That is, the measure is "zero" for the points outside the invariant set C_c and is invariant with respect to the inverse of information E (brain-body relationship).

If we consider an example of a map whose invariant set is Cantor's middle third set we will find that all our conditions for embryonic development are satisfied.

Let C be the real number line and let

$$E(c) = (3/2)(1 - |2c-1|). \tag{6-6}$$

In this case E is a two-to-one map of C unto itself. Dynamical activities of the embryology in cytoplasm (particularly in the *cytoskeleton*) possess some sort of *random perturbation*, which means that it is necessary to consider the "random" Cantor set with properties such that the Hausdorff dimensions of the final perturbated set C_c has the probability equal 1. Accordingly the model evolution (E_v) and adaptation (A) is controlled by cytoplasm through the nucleus (DNA) as a "feedforward"-"self-similarity"-"feedback" system.

Solution for $f[E,E_v]$ $(t_1^\alpha + t_2^\alpha + ...)$ as an expected value of the sum of the "a" powers of the scaling relations is (Falconer, 1990):

$$f[E,E_v] \; (t_1^\alpha + t_2^\alpha + ...) = 1. \tag{6-7}$$

Let embryology (E) and evaluation (E_v) be randomly perturbed systems as an example of randomly perturbed Cantor subsets of [0,1] as follows: First, let E be random by the uniform distribution on [0,1]. Then choose Ev, between E and 1, as random according to the uniform distribution on [E,1]. There are two intervals $C_1 = [0,E]$ and $C_2 = [E_v,1]$. The next step is similar to the previous one, which means that each interval has two subintervals: C_1 has $C_1[1_F]$ and $C_1[2_M]$, and C_2 has $C_2[1_F]$ and $C_2[2_M]$ (F-female, M-male).

$$\int_0^1 [E^\alpha + 1/(1-E) \int_E^1 (1-E_v)^\alpha dEv] \; dE = 1. \tag{6-8}$$

Equation (6-8) is satisfied if and only if

$$\alpha = [\sqrt{5}-1]/2$$

which is the *golden mean* ($\alpha = \phi$).

This spatio-temporal law from zygote trough embryology to organism in evolution we will call K_∞ exactly logic.

The second spatio-temporal logic exist on molecular level, as information mapping from DNA through RNA to proteins. Proteins are other side of genetic code and each protein possess molecular code for cooperativity. It was shown for hemoglobin that molecular code for cooperativity arise from "symmetry" role "concerted" quatenary switching and "sequential" modulation of binding within each quatenary form (Ackers et al, 1992). For biomolecules as monomers code for cooperativity with other biomolecules could arise from "symmetry" role, secondary structure, and "sequential" modulation of tertiary structure by charge (+ and -) distribution. There are *three* amino acids (lysine, arginine and histidine) with *positive* charge, and *two* amino acids (aspartic and glutamic) with *negative* charge. Their ratio is 3/2 and can satisfied Equation 6-6. Charge ratio, charge distribution and switching mechanism determinate main properties of protein dynamics. As well known tubulin proteins are major component of centrosome (Dustin, 1984). It was shown that molecular code for cooperativity of tubulin arises from symmetry properties., charge switching in subunit, GTP (energy source) and quaternary switching (between alpha and beta tubulin subunits) [10]. This tubulin molecular code for cooperativity based on "discrete" properties (switching) can be written in form:

$$2^{32 + \frac{HGTP}{2} \log_2 32} = 10^{10} \qquad (6\text{-}9)$$

Molecular code for cooperativity of tubulin as energy state of thermal, electrical and elastic properties (energy state as "continuum") has the same value (10^{10}), because there are 10 elements (ϵ_1, ϵ_2, ϵ_3, ϵ_4, ϵ_5, ϵ_6, D_1, D_2, D_3, and ΔS) each possessing 10 coefficients and they can influence each other while at the same time one element of ten may be repeated more than once. Equation 6-9 is based on tubulin structures which could smoothly approximate properties of continuum as energy state of matter. In this case we have smoothly approximation Cantor "continuum hypothesis"

$$2^{\aleph_0} = c \qquad (6\text{-}10)$$

in finite biological structures because: (1) exponent of Equation 6-9 represents "power" of set of all possible ways of "switching" in proteins (based on the ratio 3/2 - *positive/negative* charge), and (2) value 10^{10} represents "power" of set of all possible ways of states of proteins as "continuum" matter.

Spatio-temporal logic based on Equation 6-9 (as smooth approximation of Equation 6-10) we will call K^2_∞ exact logic. Through Equations 6-2b, 6-5, 6-6 and 6-8, both K^1_∞ and K^2_∞ logic have same categorical logical structures of $\aleph_0$ (aleph zero).

6.5.3 Mathematical Basis of the Aleph-Zero

Aleph zero ($\aleph_0$) is the first transfinite number invented by Cantor more than one hundred years ago (Dauben, 1979). His new number deserved something unique in mathematics, particularly in set theory. He didn't wish to propose a new symbol himself, so he chose the aleph, the first letter of the Hebrew alphabet. As the pivotal element between the finite and transfinite domains, aleph zero ($\aleph_0$) we use to explain fundamental information processes in biology.

Some elementary notation of aleph zero algebra are:

1. $\aleph_0 + 1 = \aleph_0$
2. $\aleph_0 + \aleph_0 + \ldots = \aleph_0$
3. $\aleph_0 \cdot \aleph_0 = \aleph_0$
4. $n^\aleph = c \ (n = 2, 3 \ldots)$
5. $\aleph_0^\aleph = c.$

6.5.4 Aleph-Zero Logic

Tarski and Lukasiewicz considered many-valued systems of logic (Lukasiewicz, 1970). Lukasiewicz first extended classical (two valued) logic into three-valued system. He also proposed infinite-valued logic (L_∞) whose truth values are taken from the set S of all rational numbers in the unit interval [0,1]. There are various ways to do it but we have proposed one possible way based on *random perturbation*. Today there are three classes of finite non-logical vocabularies based on aleph zero categorical theories (Weaver, 1988). The classification is based on looking at the number of bound variables which are necessary to isolated complete types. In our future work we will propose in detail a classification based on smooth approximation of finite simple mathematical groups.

We propose a new logic K_∞ based on *aleph zero* system and random perturbation (Koruga, Hameroff, 1992). K_∞ logics based on two sub-logics K_∞^1 and K_∞^2 which go together as one. Finally we have a system of *two* (K_∞^1 and K_∞^2) which is *one* (K_∞) through *three* (K_∞, K_∞^1, K_∞^2). The scientific motivation for this approach is found in biology. It is possible to make a link between K_∞ logic and others because Boolean, fuzzy and infinite valued logics are based on the set theory. They are related to each other, as are set theories are (ordinary, fuzzy and transfinite). Similar to rule relationships in logic, neurocomputing will be based on *three* main paradigms: *ordinary*, *fuzzy* and *aleph*.

6.6   Quantum Theory and the Cytoskeleton

A quantum mechanical view of reality, in which probabilistic rather than deterministic behaviors occur, may be relevant to the brain/mind. "Perhaps our minds are qualities rooted in some strange and wonderful feature of those physical laws which actually govern the world we inhabit" (Penrose, 1989). Thus, despite the fact that quantum theorists differ greatly in their interpretations and that quantum theory itself may be a "stop-gap . . . until science gives us a more profound understanding of Nature" (Penrose, 1989), quantum effects may be relevant to the brain/mind.

Where and how could quantum effects (e.g. complementarity, indeterminacy of states, wave/particle duality, non-locality, acausality, observer effects on collapse of wave functions, etc.) interact with the brain mind? Eccles (1992) has proposed that quantum indeterminacy results in probabilistic, acausal neurotransmitter vesicle release from pre-synaptic boutons, and Penrose (1989) has suggested quantum effects on neuronal function. However, the coherent structure and molecular, nanoscale dimensions of cytoskeletal

F:6-18

RELATIVE POWER OF THREE MAIN LOGICS

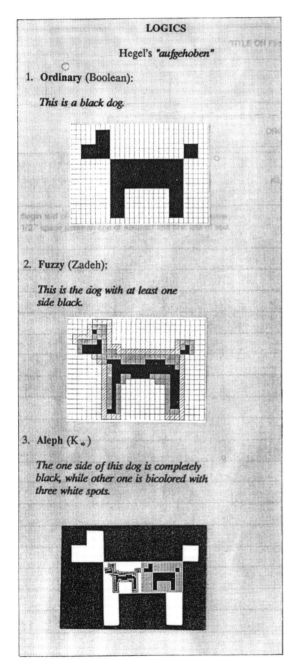

Today information technology is based on Boolean logic, as values 0 or 1 (white or black). Based on this logic, only one of these two states is possible. We recognize only one side of the dog, which we see, and we don't know that another side exists.

Another logic which is more powerful than Boolean is fuzzy, because there are more interstates between black and white. Using this logic, we know that another side of the dog exists, but we cannot say anything about it.

A new logic ("Aleph"), which we propose, is based upon "genetic and embriological knowledge," and we know not only that the other side of the dog exists but that the other side is bicolored with an exact number of spots.

components such as MT may be most appropriate for quantum level effects. While such effects could occur at larger and smaller scale levels of brain organization, it is the level of the cytoskeleton at which nonlinear, non-equilibrium dissipative dynamics occurs, single electron (or proton) events can couple to protein conformational states, coherent excitations can cooperatively lead to long-range order, and symmetry is most relevant.

As described by Insinna (Hameroff et al, 1993), particular features of a device needed to detect spontaneous, parallel occurring quantum transitions and to act as a ground element for a parallel-to-serial conversion (non-local, acausal-to-local, deterministic) may be defined:

1)   The device should possess the capability of amplifying microscopical changes or transitions occurring at the quantum level and of displaying them on a macroscopical scale. This could involve either only a few components with many degrees of freedom or energetic states, or numerous components with limited states or energetic levels (for instance, binary). The device should obey the laws of entropy; its components should, for example, be in continuous uncoordinated motion or randomly oscillating before the occurrence of the synchronistic event or state-vector reduction.

2)   After the occurrence of the event, the device (or system) should display discrete meaningful patterns or meaningful groupings of its components, either through their positions or their energetic states, each component or state carrying a "bit" of information. The "emerging" patterns should be coherent and stable; they should exclude incidental or transient random coherence.

3)   The global features of the device/system, for instance its physical constraints, should contribute to limiting the number of possible patterns. This means that the system should be a limiting factor in order to set forth a finite number of possibilities and enable the translation (recognition) of the emerging coherent patterns into meaningful results.

Such devices occur as nonequilibrium dissipative systems, and such systems are described by Prigogine, Haken, and Fröhlich. Prigogine and Haken demonstrated that macroscopic coherent behavior can emerge from micro-chaos in a system working far from thermodynamic equilibrium, that is progressing from order to disorder. The emergence of dissipative structures is the result of random microscopic fluctuations occurring in the system. Because of inherent non-linearities, these fluctuations are not dampened, but are amplified and impose a new order upon the whole system.

Descriptions of dissipative structures in living systems have largely focused on readily observable reaction diffusion patterns such as metabolic oscillations of ATP concentrations in the glycolytic cycle (Goldbeter and Nicolis, 1976; Goldbeter and Caplan, 1976), wavelike aggregation of the slime mold Dyctostelium discoideum being regulated by cAMP oscillations (Goldbeter and Segel, 1977), and regulation of enzymatic activity for glucose metabolism in the bacterium Escherichia coli depending on the activity of the lac operon (part of the bacterial genome) (Nicolis and Prigogine, 1977). In eukaryotic cells, the existence of a continuous biochemical oscillator regulating and synchronizing cellular division has been proposed (Kauffman and Wille, 1977; Nicolis and Prigogine, 1977), although cell cycles and clocking may be regulated by cytoskeletal vibrations (e.g. Puiseux-Dao, 1984). Microtubules can generate and manifest dissipative patterns (Tabony & Job, 1992).

Another description of emergence of coherence is Fröhlich's suggestion that biological molecules, if maintained in a nonequilibrium state at constant temperature and steady external energy supply, would start behaving in coherent fashion under the effect of longitudinal frequency modes (Fröhlich, 1968, 1970, 1975a, 1975b). Through strong excitation, some polar groups of these molecules would be stretched, producing large dipole moments and entering a metastable state comparable to that of dipoles in ferromagnetic materials. If further excited, the oscillating dipoles would condense into the lowest energy state, displaying coherent oscillations of a single mode as well as long-range interactions and correlations. Water molecules bound at biomolecular surfaces would also oscillate coherently and cooperatively. Vitiello (1992) and Jibu and Yasue (1992) show that such ordered water on cytoskeletal surfaces provides symmetry in cytoplasm which, when perturbed, can yield quantum phenomena such as bosons, vibrational phonon quanta suitable for signaling, and long-range effects.

Prigogine's, Haken's, and Fröhlich's nonequilibrium structures, though in different systems, are similar phenomena. In these systems, chance (probabilities) and deterministic laws coexist, each of them contributing in an equal manner to the emergence of new complex order and of global coherence. Peat (1988) has suggested examining such systems as probable substrates for synchronistic (non-local) quantum events. Coherently oscillating cytoskeletal networks and microtubules may be dissipative systems suitable for quantum detection, amplification, and stabilization.

Causality is founded on sequential events, probably because our brains work mostly in serial fashion when it comes to comprehension of events. Perhaps our brains (in particular, coherently oscillating cytoskeletal microtubules) discriminate, detect, and serialize synchronistic and non-local, parallelly-distributed discrete quantum states which spontaneously and simultaneously emerge (from what Jung referred to as the "collective unconscious"). This assertion is supported by facts such as short dreams containing such enormous quantities of information that hours are usually needed in order to put it into serial description. Often we don't know which is the beginning or the end of a dream, possibly because of a parallel-to-serial conversion mechanism existing in our conscious processes. The same thing happens with intuitive ideas, which are "received" in a short glimpse and whose interpretation sometimes needs days to be correctly, or serially, ordered. Comprehension of intuitive ideas seems to be a genuine sequential process of our psyche. It enables us to transpose synchronistic unconscious and parallelly received events into finite space-time coordinates, establish value scales and coherently order the received data according to some basic logic (causal and sequential) rules. Again, the parallel, non-local emergence of information and the serial process we suggest here complies with the new holonomic (non-local and holistic) theories of the brain such as the one suggested by Pribram (1991).

One global feature of a dissipative system capable of quantum detection is limiting the number of possible patterns, enabling recognition of meaningful results. Insinna (1992) has suggested that the possible states of MT dynamical systems ("basins of attraction") correspond with archetypal images as described by Jung. Wuensche (1992) shows that basins of attraction for cellular automata can be calculated and represented (Figure 6-16) and suggests that such attractors, which exist in multidimensional state space, are the cognitive substrate of the automata ("the ghost in the machine"), in that they represent their catalogue of behaviors.   Perhaps the mind/brain relationship may be considered analogous to that of the attractor/automata. If so, the mind may be considered to exist in state space (or some high-dimensional space) comparable to "World 2" proposed by Eccles (1986; Popper & Eccles, 1977).

Another quantum aspect is that of wave interference. Pribram (1972; 1991) considers interference of coherent waves as a substrate for consciousness ("holography"). Coherent excitations of MT and their subunits may result in coherent waves (e.g. coupled to calcium ions causing sol-gel transformations) in the cytoplasm (Figure 6-17). Such coherent calcium coupled waves have been imaged in living cells (Lechleiter et al, 1991). Holographic imaging may result and represent "agent" or cognitive imagery in cells, including neurons, and, collectively, the brain/mind.

6.7   The Cell's Information Invariants

6.7.1   DNA

One nucleotide of DNA is composed of three elements: a base, ribose and a phosphate group. Four types of bases may be represented: adenine, thymine, guanine and cytosine. Only two types of bases actually exist, these being a purine base and a pyrimidine base. Purine bases are adenine and guanine, while the pyramidine bases are thymine and cytosine. Nucleotides are interconnected by hydrogen bonds organizing them in a specific double-helix structure. From the aspect of organization of structure one such double-helix is an aperiodic crystal (Schrodinger, 1967). The term "aperiodic" signifies the irregular interchange of bases inside the helix while the phosphates and ribose are located on the outside making up a periodic crystal structure. The irregular repetition of the bases within the helix represent properties of living beings which, from the information point of view, manifest a code system. As is well known, in the genetic code, one triplet of bases codes one amino acid. The basic genetic code is coded by 20 amino acids and there exists a "stop" as three more codons. On the basis of this result we can conclude that there exist 61 codons which code 20 amino acids. The genetic code from the aspect of biochemistry is based on a triplet, which in combination of four bascs gives a total of $4^3 = 64$ possible combinations.

The mechanism of protein synthesis is well-known. Messenger RNA (mRNA) is synthesized from one end of the DNA double helix, while the other end of the helix remains in the nucleus, making possible the synthesis of another chain of DNA. The complete genetic information is preserved and remains inside the nucleus. From mRNA through carrier RNA (tRNA) to ribosomal RNA (rRNA), there is a continual transmission of the genetic information message making in effect proteins the "other side" of the genetic code.

6.7.2   Tubulin

The primary structure of tubulin extracted from porcine brain is well-known (Dustin, 1984). Secondary structure of tubulin has been studied implementing the circular dichroism (CD) experimental method (Ventila et al., 1972) and by Raman spectroscopy (Simić-Krstić, 1991b; Audenaert et al, 1989), as well as the theoretical method of Chou-Fasman (Fasman, 1989). In its depolymerized state the secondary structure of tubulin is 26% α-helix, 41% β-sheet and 33% random coil; while in its polymerized state microtubules are 40% α-helix, 31% β-sheet and 29% random coil. Utilizing the Chou-Fasman mode, for α-tubulin it was possible to predict the existence of four α-helix, four β-sheet and two random coils, while β-tubulin has six α-helix, one β-sheet and seven random coils. α and β tubulin subunits, according to the model of Mandelkow (Mandelkow & Mandelkow, 1989), are able to bond as α-β heterodimers with the aid of the strong binding GTP. The bonds of the dimers in the protofilament are established with the aid of GTP, but intersubunit bonds are weaker so that they can freely interact. Because the X-ray crystal structure of tubulin is not available, there is no straightforward way to relate the above observations to the three-dimensional folding of the polypeptide chain (Mandelkow, 1985).

6.7.3 Microtubules

As previously described, however, tubulin plays a role in the process of learning (e.g. Mileusnić, 1980), and careful study of these structures also shows that they possess a code system (Koruga, 1986). That code system is structurally a binary system and has the characteristic 2^6 with a distance of 5. When there is a biophysical transfer of information from one point to the other in the MT, the binary coding system of microtubules becomes terniary $K_2 = 3^4$, because the hexagonal surface organization transforms into a new cubic state, making possible the propagation of information. The basic informational unit of the microtubule is the region bounded by the microtubule associated proteins (MAP), and relationship between K_1 and K_2 codes, as are shown in Figure 5.3-1.

Structures such as DNA and microtubules possess code systems; coding law may be applied to study their information principles. The equivalent of the coding law for microtubules are the laws of unit sphere packing (Leech & Sloane, 1971; Thompson, 1983; Sloane, 1984). The $Oh(6/4)$ symmetry group for packing of unit spheres which are equal to unity is most appropriate for microtubules.

We will also take (1) the point symmetry groups--from crystallography-- and (2) the limit groups of symmetry--Curie groups (Shaskolskaya, 1982)--in order to connect energy and structure from the information point of view. As information law is logarithmic in character (Hamming, 1986) we will use the operator L as the operator of the information code. We are thus searching for the information invariant which should be provided by their coding systems.

Microtubules possess two structural coding systems (Figure 5.3-1) which depend on the unity of structure, energy and information based on limit symmetry groups known as Curie's groups (Shaskolskaya, 1982). Thus, every tubulin subunit, both α and β, consists of a Curie group $\infty \infty m$ to which are attached all physical properties of the 32 crystallographic groups. Consequently, depending on the position of the subunit inside the microtubule, energetic processes will activate one or more of 13 symmetry groups responsible for the magnetic properties of microtubules. The relationships among 32 crystallographic groups and Curie's groups on the basis of their energetic properties are presented in Figure 2-17.

Curie's symmetry group made up of 13 elements is responsible for the magnetic properties, and the group made up of 10 elements is responsible for the electric properties of microtubules. Because cilia, flagella and centrioles have an organization of 13 + 10 microtubules, protofilaments consisting of 13 subunits have magnetism as their primary state, while tubules with 10 protofilaments have electric characteristics as their main properties. However, in these organelles there exists a complete unity of electromagnetic properties. Figure 2-17 illustrates the main relationship among the 32 crystallographic groups and Curie's groups in microtubules, as well as the shape of Curie's groups, which gives the matrix of 10 elements with 10 coefficients each for $\infty \infty m$, ∞ / m and ∞m Curie symmetry groups.

On the basis of Figure 5.3-1, in microtubules with 13 protofilaments, each dimer has the value of the number of protofilaments in the microtubules so that the distance between the subunits is $d = 5$. As this is the result of the coding system, there must exist a direct relationship between the information characteristics of the dimer and the information characteristics of the microtubule on the basis of an information invariant as a space-time entity. To that end should be considered the information characteristics of the dimer, based on energy properties of symmetry groups.

6.7.4   Information Invariant of α-β Tubulin Dimer

Crystallization of tubulin subunits has been impossible and X-rays of tubulin crystal structures have been unavailable (Mandelkow, 1985) because the basic crystallographic state of tubulin is given as a monomer given in the symmetry $\infty \infty m$ (limit symmetry group), and not by one of the 32 basic crystallographic point symmetry groups as is the case with most biological and non-biological materials. From this aspect tubulin presents a unique protein structure in which the genetic code (based on the characteristics of an aperiodic crystal) transforms through amino acids into a new crystal state, ordained by the $\infty \infty m$ characteristics of the Curie symmetry group. This new crystal state is cardinal in contrast to the periodic (point symmetry groups) and aperiodic (repetitive structures on the outside, non-repetitive on the inside--but both as point symmetry groups) crystal state because it contains all 32 point symmetry groups.

It is known from crystallography (Nye, 1957; Sirotin & Shaskolsky, 1979) that the matrix of structural energy characteristics with the ultimate symmetry $\infty \infty m$ have the following meanings: ϵ_1, ϵ_2, . . ., ϵ_6 show representations of factors of conformational change which take place under the influence of temperature; δ_1, δ_2, . . ., δ_6 represent changes of stress in the material due to conformational changes; D, electric induction; ΔS, entropy; and ΔT, absolute temperature per unit volume. On the basis of the matrix, it is possible to realize 10^{10} different occurrences, because there are 10 elements each possessing 10 coefficients and they influence each other, while at the same time one element of 10 may be repeated more than once. Considering the new classification of amino acids (Sneddon et al, 1988) which are carbon-based natural semiconductors, it is possible to realize off-on functions in the tubulin dimer on the principle of n-type (aspartic and glutamic acids) and p-type (lysine, arginine and histidine) switches incorporating five amino acids and giving 2^5 possibilities (Koruga & Simić-Krstić, 1990).

The tubulin subunit α, as a structure possessing $\infty \infty m$ symmetry, interacts with the other subunit β, which also possesses the symmetry $\infty \infty m$. As the symmetry group is composed of 32 members, the same 2^5, this means that the dimer will give 2^{32} or 2^{25} possible occurrences. Because the tensor of dielectric permittivity has the symmetry $\infty \infty m$ (Shaskolskaya, 1982), the energy state during n-type and p-type switches, and not the structural properties of tubulin, will give 2^{32} possible occurrences. However, with the formation of the dimer it is possible to obtain one more hybrid member which is ordained by the characteristic of the bond in the dimer. The bond in the α-β dimer requires GTP so that the energy state of the dimer under the influence of the new member (2^{32}) will be characterized by the energy potential:

$$GTP^{4-} + H_2O \rightarrow GDP^{3-} + HPO^{2-}_4 + H^+ \ (0.49eV) \qquad \text{(6-11)}$$

This means that two subunits will share H^+ (0.49 eV) and that they will form a symmetry group which is a hybrid state of two subunits of 32 symmetry groups, but such that the odds of influence from each of the 32 symmetry groups is $\infty \infty m$ on the formation of a new group. We can write the following on the basis of the tubulin crystal properties information measure:

$$\mathcal{H}^{dim} = 2^{32 + (\mathcal{H}+/2) \cdot \log_2 32} = 10^{10} \qquad \text{(6-12)}$$

where $\mathcal{H}^+ = 0.4877$ eV. ($\mathcal{H}^+ = H^+ - \mathcal{H}^\delta_{H2O}$ in reaction $H^+ + OH^- \rightarrow H_2O^*$ where water has a spatial fivefold symmetry--screw and/or icosahedral).

The result of this is that the hybrid state of the dimer in microtubules must have a shape defined by the Curie symmetry groups which ordain the pyroelectric effect on the surface of the dimer, as a result of the symmetry laws. On the basis of these laws (Sirotin & Shaskolsky, 1979; Shaskolskaya, 1982) the pyroelectric effect of the dimer on the surface will give a specific shape based on limit symmetry groups. From this figure we see that one dimer possesses a symmetric and an asymmetric component, a result of properties of magnetic symmetry of the point symmetry group which give $\infty \infty m$ limit symmetry group. This means that the shape of the dimer in the protofilament of 13 subunits which possesses the primary magnetic properties of the Curie group would look as in Figure 2-17.

Using contemporary techniques such as scanning tunnelling microscopy (STM), it is possible to study the surface state of materials (Stobbs, 1989), including biological materials (Amrein et al, 1988; Simić-Krstić et al, 1989). Research carried out utilizing STM on microtubules isolated from porcine brain (Hameroff et al, 1990; Simić-Krstić, 1991b) shows that the surface shape of the dimer of tubulin is the same as that obtained on the basis of crystallographic laws of ∞m Curie symmetry and Equation 6-12.

6.7.5   The Information Invariant of Microtubules

If this value is the information invariant of the microtubule code, it must be satisfied from the aspect of structure because the value of the dimer is equal to the value of the microtubule--the number of protofilaments of the microtubule. As we have stated earlier, microtubules possess two apparent coding systems: one is $K_1 = 2^6$ and the other is $K_2 = 3^4$. If we take a logarithm as the basis of information laws and apply a logarithm operator: $L[L(a^b) + L(c^d) = bL_a a + dL_c c = (b + d) L_{(a + c)} (a+c)$ *if and only if* $(b + d) = (a + c)$. For 10^{10} we can write the relationship as:

$$6L_2 2 + 4L_3 3 < 10L_{10} 10 \qquad (6\text{-}13)$$

From Equation (6-13) we can see that it does not satisfy the information invariant, the information value obtained in observing the dimer. Because MT are self-assembly entities and the dimer value is the same as the MT value (Figure 5.3-1), there has to be a third member to fill the information value up to $10.^{10}$ Because of this we must write

$$6L_2 2 + 4L_3 3 + 0L_5 5 = 10L_{10} 10 \qquad (6\text{-}14)$$

On the basis of this equation, we see that microtubules must possess a third code system $K_3 = 5^0$. This is understandable because dimers are a part of microtubules and dimers are phosphorylated by exchangeable GTP (Mandelkow, 1985). This implies that this information code is associated with Equation 6-12 and that the energy level of the hydrogen ion influences water molecule orientation through dimer activities with $\log_2 32$ ($\mathcal{H}_\xi = H^+ - \mathcal{H}^+$). The position of the subunit, i.e. the protofilament, will ordain which of the 13 of a total of 32 symmetry groups activates according to the laws of crystallography and electron density for each subunit or protofilament.

6.7.6   DNA Information Invariant

Proteins, including tubulin, are the result of the synthesis of amino acids, and are the "map," the other side of the genetic code. Consequently, the information invariant of the gene structure code must also possess an information value of 10^{10}. Only in this way can proteins as nanoscale devices carry out functions which coded in DNA (mapping information from DNA to protein).

Structures which make up genes are four bases. These same bases can make other types of structures having information characteristics, for example cyclic monophosphates. Of these four structures, only one plays an important role in bioinformation processes: cyclic adenine monophosphate (cAMP), which is the product of adenyl cyclase. In that case, besides the coding system 3^4, another coding system is evident: 4^1. On this basis it is possible to write

$$3L_44 + 1L_44 < L_{10}10 \qquad (6\text{-}15)$$

In Equation (6-15) we see that again there is a missing coding system which should fulfill the information value of the information invariant. This must be the binary coding system 2^6. On the basis of this we can write

$$3L_44 + 1L_44 + 6L_22 = 10L_{10}10 \qquad (6\text{-}16)$$

Because the basis of information laws is logarithmic, a code given in triplet can have eight basic information states in binary code [2^3], repeating cyclically, because $\log_2 8 = 3$.

As Equations (6-14) and (6-16) have the same information value and both have the characteristic of the invariant, it is clear that by bringing in the third member of Equation (6-16) can explain the coding system of DNA and the number of amino acids which codify the genetic code.

The binary characteristic of the genetic code may be the factor which determines the number of amino acids which can be coded with the 61 codons (Swanson, 1984; Rakočević, 1988).

6.7.7 Cell's Information Invariant

The information value as the given invariant in the relation DNA → tubulin → microtubules → DNA' is tripled, and that the information invariant as the space-time pattern-entity is given the code value 3×10^{10}. This information invariant can be written as $\mathcal{H}_k = 3 \times 10^{10}$ and can be called the cardinal biological information invariant (cBii) as left-right unity.

(1)   Considering the property of the tubular dimer to have an information value consistent with the value of the information invariant, and the property of microtubules to have the same information value as a dimer, these

structures have a *fractal-like* relationship of the **part** to the **whole**. Also shown is that the information invariant as a space-time entity is important for the "development" of intelligent hardware. Intelligent hardware on the molecular level creates new intelligent hardware in the process of cellular differentiation which can interact with its environment. During embryological development, starting with the insemination of the egg to the development of the organism as a whole, there exists a complete coordination in space and time, and this information as a space-time entity through the information variant leads to the organism as a whole. This process of differentiation almost always comes to the same end. As this information invariant is a completely coordinated space-time structure, $\mathcal{H}_{\mathcal{H}} = 3 \times 10^{10}$ has a space-time dimension. The *space* dimension in this case, can be taken as *centimeter* (cm), the *time* dimension as *second*(s). We then get $\mathcal{H}_{\mathcal{R}} = 3 \times 10^{10} \; [cm \; s]$ as a calm state of space-time, as information storage. We thus see that the basic information invariant is equal in value to the space-time invariant written as a change in space-time, the speed of light ($c = 3 \times 10^{10}$ cm s^{-1}). In other words, the "intelligent hardware" based on nucleic acids (DNA) and proteins (MT) as a cognitive space-time entity in *water* is the equivalent of the space-time invariant based on the change of space-time through the speed of electromagnetic waves (Koruga, 1992).

Molecular computing, as a link between physics and biology (Conrad, 1984), is the key to understanding how information is processed in the cell, as well as in tissue, organ and organism embryogenesis. Without a solid understanding of these processes, it is not possible to fully understand the workings of the brain and mind. Linking of the brain and the mind can be considered through the INFON, as a 3×10^{10} entity, which couples biological matter and electromagnetic waves, as both are of the same space-time invariant, produces a new entity, the *Link*. This new entity, based on electro-magnetic values of biological structure and biochemistry of structures which give and support the existence of *cBii*, permits a whole new phenomena--*cellular awareness*--as a preamble to the consciousness of human beings based on the parallel actions of 10^{10}-10^{12} neurons. For advancement in computer sciences and technology, understanding these mechanisms has a primary importance, especially the consequential evolution, based on the *INFON*, of the interrelations of Nature-Brain-Mind-Computer.

CHAPTER 7

**FROM NANOBIOLOGY
TO NANOTECHNOLOGY:**
Basic Concept

SELF-ASSEMBLY

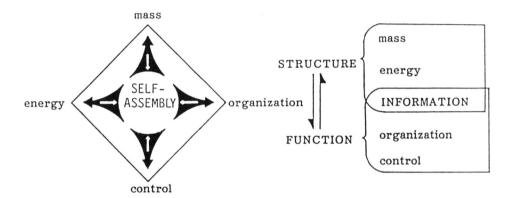

A schematic view of the relation between the main entities involved in forming structure and function. Information as a space−time entity is the result of the processes which enable life to exist.

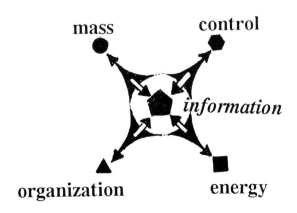

7.1 Self-Assembly

There are different approaches to defining self-assembly, but in the context of Fuller's terminology: *self-assembly is synergy of information*. According to Fuller, "Synergy is behavior of integral, aggregate, whole system unpredicted by behaviors of any of their components or subassemblies of their components taken separately from the whole." Information, for any physical system, is the unity of mass, energy, organization and control. The modern classical picture considers *structure* and *function*. Alternatively, mass, energy and information characterize *structure*, while information, organization and control represent *function*. As we can see (Figure 7.1-1), information is a common link between these two categories. In a novel approach, we define **information** as a *measure of mass, energy, organization and control based on symmetry*. In this definition *measure* is a relation (*immediate* and/or *mediate*) of *quantity* and *quality*. This indicates that there are different types of information.

This new approach needs to consider all four categories (mass, energy, organization and control) from a symmetry point-of-view: (1) *Mass symmetry* is the order of atoms and molecules; (2) *energy symmetry* is the space-time pattern of electronic, vibrational, rotational spectra; (3) *organization symmetry* is left-right orientation, hierarchy of systems, and timing; and (4) *control symmetry* is optimal control of the Hamiltonian process, adaptive control of the system, and intelligent control of system-environment interaction. Each self-assembled system will have an *information spectrum* as a set of values of the state and/or process of *each part* and the *whole*.

In the transition from nanobiology to nanotechnology, it is necessary to identify the major self-assembled principles and structures and to implement that knowledge in nanotechnology in a manner similar to the relationship between the flight of a bird and of aircraft: both use wings but in different ways. The essential structures of life are *nucleic acids*, *amino acids* and *water*. In Figure 7.1-2, the basic relationship of these structures is shown relative to their mass, energy and information. Structures like ATP and GTP are energy sources in the cell, but these energy structures are also basic structural components which have information capabilities (DNA, RNA, cAMP and cGMP). Generally, nucleic acids in DNA have *right-handed* orientation (there are some DNA forms with left-handed orientation, but they represent a very small percentage of the total). DNA, through RNA, provides the main spatio-temporal pattern (information) for synthesis of *amino acids* into proteins. There are a thousand proteins in each cell, but our focus is on *tubulin* as a self-assembly protein. Amino acids in proteins have *left-handed* orientation. This organization principle (left-right) is a crucial one in biology. During the process of amino acid synthesis, phosphorylation based on ATP is also important. Tubulin subunits, α and β, make microtubules (MT), which interact with cAMP kinase. MT are biological information devices based on *Golden mean* properties (Chapter 5) and are the main part of the *centriole*, a cell organelle with *left* and *right* spiral orientation. Centrioles are the main *synergetic* devices in the cell and regulate: (1) dynamics in cytoplasm; (2) the main relationship between nucleus and cytoplasm; and (3) the spatio-temporal relationship between cells during the process of embryogenesis. 70 percent of each cell is water; thus, cellular processes occur in water solution. However, much of the water is "ordered" and dynamically cooperative with cytoskeletal surfaces.

As an important example of biomolecular self-assembly, we use tubulin and present its relationship to Fullerene nanotechnology in Figure 7.1-3. There are four main levels: atom-molecule; molecule-organelle; organelle-cell; and cell-organism. Self-assembly mechanisms are similar but not the same at each level. Fullerene self-assembly nanotechnology has two branches in symmetry transformation, while biology has only one way. In Figure 7.1-3 we propose a main pathway for a self-assembled molecular computer based on Fullerene C_{60} as a basic principle, from nanobiology to nanotechnology.

At the first level (atom-molecule; Figure 7.1-3) both biological and Fullerene systems have self-assembly processes. In biology, the *functional solvent* is water, while in Fullerene systems it is helium. At the second level (molecule-organelle), the biological approach begins with a pool of GTP-T_u subunits in water, while in the Fullerene system, it will be a pool of C_{60} in an electromagnetic field of special liquid-crystal functional solvent. The symmetry potential in the biological (S^{biol}_{poten}) example is an enormous one (from $2\infty^3$ to 9-fold symmetry. This potential has a similar pathway to ATP$\rightarrow$ADP$\rightarrow$AMP (three steps: I, II and III). In the case of Fullerene, there are also three steps but with two overlapping pathways. Symmetry potential (S^{Full}_{poten}) also exists for Fullerenes but is much smaller than in biology. Thus, a dimer of the C_{60} molecule has to be *complementary* to the functional solvent, which is not the case in the biological solution (organized water is a *partial energy information complement* of each protein, being the universal solvent).

F:7.1-2

SELF-ASSEMBLY OF NUCLEIC ACIDS

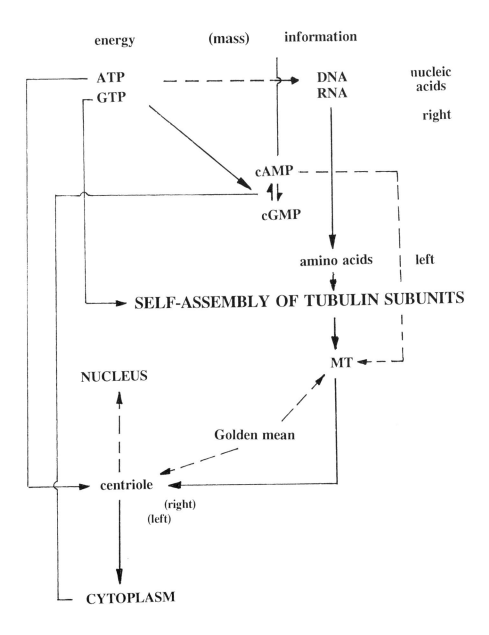

cell   70%   water

F:7.1-3

SELF-ASSEMBLED MOLECULAR COMPUTER

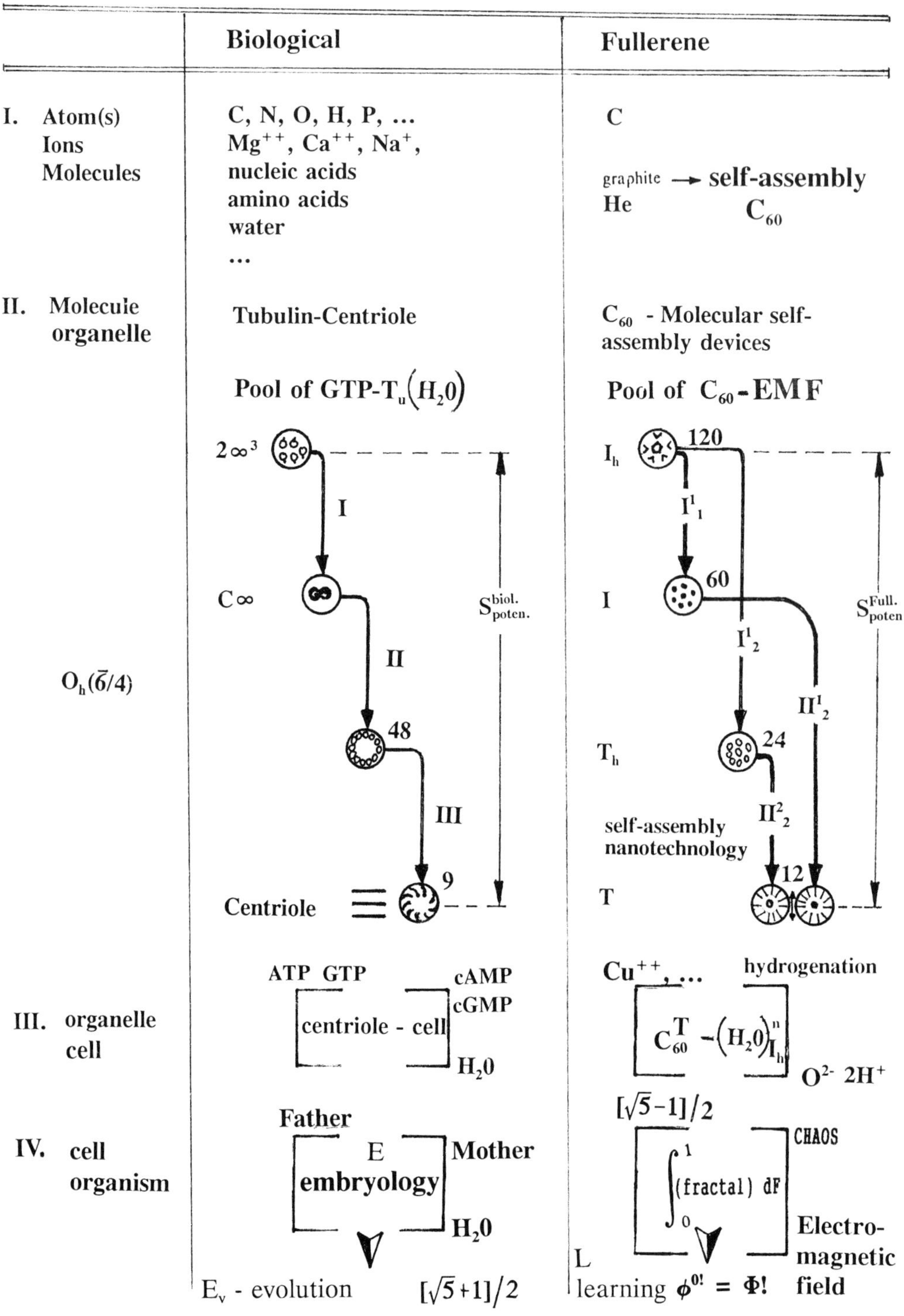

	Biological	Fullerene
I. Atom(s) **Ions** **Molecules**	C, N, O, H, P, ... Mg^{++}, Ca^{++}, Na^+, nucleic acids amino acids water ...	C graphite $\rightarrow$ **self-assembly** He C_{60}
II. Molecule **organelle**	Tubulin-Centriole Pool of GTP-$T_u(H_2O)$	C_{60} - Molecular self- assembly devices Pool of C_{60}-**EMF**

$O_h(\bar{6}/4)$

Biological column diagram:

$2\infty^3$ → I → $C\infty$ → II → 48 → III → Centriole $\equiv$ 9

$S^{biol.}_{poten.}$

Fullerene column diagram:

I_h → 120 → I^1_1 → I → 60 → I^1_2 → II^1_2 → T_h → 24 → II^2_2 → self-assembly nanotechnology → T → 12

$S^{Full.}_{poten}$

| **III. organelle**
cell | ATP GTP $\rceil$ cAMP
$\lceil$ centriole - cell $\rfloor$ cGMP
$\lfloor$ H_2O | Cu^{++}, ...   hydrogenation
$\lceil$ $C^T_{60}-(H_2O)^n_{I_h}$ $\rceil$
$\lfloor$ O^{2-} $2H^+$
$[\sqrt{5}-1]/2$ |
| **IV. cell**
organism | **Father**
$\lceil$ E $\rceil$ **Mother**
embryology
$\lfloor$ H_2O
$\bigvee$
E_v - evolution $[\sqrt{5}+1]/2$ | $\lceil$ $\int_0^1$ (fractal) dF $\rceil$ CHAOS
$\bigvee$
L
learning $\phi^{0!}=\Phi!$
Electro-
magnetic
field |

On the organelle-cell level, there are also similarities between biology and a Fullerene system, however one big difference is that the energy sources in a biological cell are ATP and GTP, while in a Fullerene nanotechnology solution it will be C_{60} hydrogenation molecules. Also, C_{60}^T devices will be *energy-information* synergetic. In the biology, MT are synergetic devices, but only from the standpoint of information (a dimer subunit has the same information "power" as an MT). In the case of the cell-organism level, fractal electromagnetic dynamics will be equivalent to *embryology*. The biological result is evolution based on the Golden Mean (GM^+), while in the Fullerene case, the result will be learning and unification of artificial and biological devices on a nanometer scale as a new *creation scenario* ($\Phi! = \phi^{0!}$).

To realize the transition from nanobiology to nanotechnology, control engineering is crucial, particularly on a molecular level.

7.2 Control Engineering and Molecular Systems

7.2.1 Introduction to Control Systems

The ultimate goal of engineering is to design and build physical systems that can perform desired tasks. Hence control systems have now become an integral part of numerous applications ranging from the simple thermostat control of the temperature of a home to the sophisticated control loops underlying the functioning of modern aircraft and spacecraft, nuclear reactors and complex chemical processes.

A typical control system may consist of several subsystems and is, in general, an interconnection or assembly of components forming a system configuration that provides a desired system response by accepting a suitably chosen input (or command) signal, as shown in Fig. 7.2-1. Thus the main objective of a control system is to control in some prescribed manner the output signal y(t) of the process (physical or biological) being controlled, by generating appropriate input signals to drive the process.

All control systems can be divided into two categories depending on the mode in which the control objective is accomplished: (i) *open loop control system* which is the simplest and the most crude form. The schematic diagram is shown in Fig. 7.2-1a, where the controller generates the actuating signal to drive the process to attain the desired output y(t). The control adjustment in such a system invariably depends on human judgement and estimate. (ii)

F:7.2-1 and 7.2-2

General Configuration of a Control System

Schematic Diagram of an Open Loop Control System

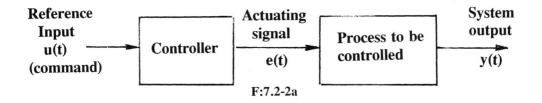

F:7.2-2a

Schematic Diagram of a Closed-Loop Control System

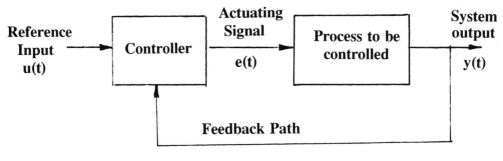

F:7.2-2b

closed-loop or *feedback control system*, which involves a more sophisticated control arrangement designed to overcome the deficiencies of open loop control systems by providing a feedback path from the output point to the input point. Thus a comparison between the actual value of the output and the desired value of output (indicated by a reference input) can be obtained and any deviation or "error" between the two signals is used to drive the process to be controlled. The schematic diagram of a closed-loop control system is shown in Fig. 7.2-2b.

Almost all of the control systems one encounters in modern applications are of the closed-loop type. An organized and systematic design of control systems typically involves the use of mathematical models for representing the dynamics of the process to be controlled and the control objectives (modelled as specifications for conducting the designs). A variety of mathematical tools ranging from ordinary calculus to more sophisticated ones such as nonlinear programming and dynamic programming have been employed in control system designs for various physical and biological processes and to realize diverse control objectives. Studies that have been directed to these during the past several decades have resulted in a vast body of knowledge and have made *control engineering* one of the most important disciplines of engineering (Dorf, 1992).

7.2.2 Complexities in Control System Designs

The problems of decision-making and control of systems operating in the real world are often complicated by several factors such as:
    (i)   complex nonlinear interconnections;
    (ii)   environmental changes and system parameter variations;
    (iii)   lack of complete information at the decision-making stations;
    (iv)   need for "optimal" performance; and
    (v)   desire for "intelligent" decision-making in the execution of control.

Due to the above, controller designs using classical (simple-minded) procedures are invariably not adequate to deliver satisfactory performance and one faces the need for more "sophisticated" design procedures. Some of these are outlined in the following.

7.2.2.1 Nonlinear Control Systems

The presence of nonlinear dynamical relations make the designs based on linear models (such as Transfer Functions) invalid. A typical example is the inertial cross-coupling phenomena in aircraft and spacecraft control problems evidenced by the equations of motion (obtained from the corresponding Euler equations) given below.

$$I_x[(\partial^2\phi)/(\partial t^2)] + (I_z - I_y)[(\partial\theta/\partial t)(\partial\psi/\partial t)] = k_1 u_1$$
$$I_y[(\partial\theta)/(\partial t)] + (I_x - I_z)[(\partial^2\phi/\partial t^2)(\partial\psi/\partial t)] = k_2 u_2 \qquad (7.2.1)$$
$$I_z(\partial\psi/\partial t) + (I_y - I_x)[(\partial\theta/\partial t)(\partial^2\phi/\partial t^2)] = k_3 u_3$$

where ϕ, θ, and ψ are, respectively, the roll, pitch, and yaw displacements, I_x, I_y, and I_z are the components of body inertia, k_1, k_2, and k_3 are the reaction-wheel constants, and u_1, u_2, and u_3 are the control signals actuating the wheels.

By defining the subsystem state vectors

$$x_1 = \begin{bmatrix} \dfrac{\partial\phi}{\partial t} \\ \\ \dfrac{\partial\phi}{\partial t} \end{bmatrix} \quad x_2 = \begin{bmatrix} \dfrac{\partial\theta}{\partial t} \\ \\ \dfrac{\partial\theta}{\partial t} \end{bmatrix} \quad x_3 = \begin{bmatrix} \dfrac{\partial\psi}{\partial t} \\ \\ \dfrac{\partial\psi}{\partial t} \end{bmatrix} \qquad (7.2.2)$$

(1) may be rewritten as an interconnected nonlinear system

$$\frac{\partial x_i}{\partial t} = A_i x_i + B_i u_i + h_i(x), \qquad i = 1,2,3 \qquad (7.2.3)$$

where

$$A_1 = A_2 = A_3 = \begin{bmatrix} O & 1 \\ O & O \end{bmatrix}, \quad B_1 = \begin{bmatrix} O \\ B_1 \end{bmatrix}$$

and

$$h_1(x) = \begin{bmatrix} O \\ -\delta_1 \, x_{22} \, x_{32} \end{bmatrix}, \quad h_2(x) = \begin{bmatrix} O \\ -\delta_2 \, x_{32} \, x_{12} \end{bmatrix},$$

$$h_3(x) = \begin{bmatrix} O \\ -\delta_3 \ x_{12} \ x_{22} \end{bmatrix}$$

where

$$\beta_1 = k_1/I_x, \quad \beta_2 = k_2/I_y, \quad \beta_3 = k_3/I_z$$

and

$$\delta_1 = (I_z - I_y)/I_x, \quad \delta_2 = (I_x - I_z)/I_y, \quad \delta_3 = (I_y - I_x)/I_z.$$

The controller to be designed for a desired performance objective will also be a nonlinear function of the state in order to account for the presence of the nonlinear interconnections. A typical controller designed for this system is as given below (more details on this system and the controller design can be found in Sundareshan (1977a)). For a set of specific values of the inertia components,

$$I_x = 14656 \ kgm^2, I_y = 91772 \ kgm^2, \quad \text{and} \quad I_z = 95027 \ kgm^2, \quad \text{and typical}$$

reaction-wheel constants $k_1 = k_2 = k_3 = 12.57 \times 10^5 \ N.m/rad$, the control signals for exponential stabilization of the overall system have been computed as

$$u_1 = -0.978x_{11} - 1.076x_{12} + 0.0026x_{22}x_{32}$$

$$u_2 = -0.981x_{21} - 1.172x_{22} - 0.064x_{12}x_{32}$$

$$u_3 = -1.052x_{31} - 1.097x_{32} + 0.0613x_{12}x_{22}.$$

For a more detailed and extensive coverage of the complexities involved in the design of controllers for nonlinear dynamical processes, one may refer to Slotine and Li (1991).

7.2.2.2 Adaptive Control Systems

The objective of adaptive control is to obtain good closed loop performance when the plant parameters and disturbances are unknown or time-varying. The system parameters may be time-varying due to environmental changes or the use of a linearized model for a nonlinear system. In the latter case, changes in the linear model parameters correspond to changes in the operating point of the system. These perturbations in the plant model parameters can be considered to be parameter disturbances. Thus the goal of adaptive control is to compensate for these parameter disturbances. This contrasts with the goal of conventional feedback/feedforward control, which is to suppress the effect of disturbances on the controlled variable only.

The basic premise upon which all adaptive controllers are designed is that for any possible value of plant parameters and disturbances, there exists a controller of fixed structure which yields the desired closed-loop performance. Thus, all adaptive control schemes are based on a conventional feedback controller which would be used if the plant dynamics were known and time invariant.

The general structure of adaptive controllers is shown in Fig. 7.2-3a. Some criterion of system performance is measured and compared with the desired performance. The deviation is input to an adaptation mechanism which modulates the controller parameters. The adjustment mechanism consists of a recursive parameter adaptation algorithm (PAA), which is generally some form of recursive least squares (RLS) estimation.

The adaptation may be done indirectly or directly. Indirect adaptation is a two-step process. In the first step, a PAA uses current and past input/output data to estimate the plant model parameters explicitly. The most recent plant parameter estimates are used in the second step to calculate the control parameters. This process is repeated every sampling period. Applicability to both minimum and non-minimum phase systems and allowing many estimator/controller combinations are advantages of the indirect scheme.

In the direct approach, the PAA adjusts the control parameters directly. The plant model is rewritten to include the controller parameters. The controller parameters are then updated by the PAA as part of the plant estimation. Fewer computations are involved in the direct scheme, but in general, direct schemes can only be applied to minimum phase systems.

Adaptive control theory has been developed from two different points of view--model reference adaptive control (MRAC) and self-tuning controllers

(STC). In MRAC, the desired closed-loop performance is stated in terms of an explicit reference model. The reference model may be a transfer function or a difference equation. As shown in Fig. 7.2-3b, the difference between the actual system output and the reference model (desired) output is input to the PAA. The design goal is to synthesize the PAA such that

$$\lim_{t \to \infty} \left[y(t) - y_m(t) \right] = 0 \qquad (7.2.4)$$

This is achieved by cancellation of the open-loop poles and zeros of the plant. This can only be done for minimum phase plants. Indirect MRAC schemes can be used for nonminimum phase plants where it is desired to replace only the poles and stable zeros of the plant. More details on the design of MRAC schemes can be found on Narendra and Annaswamy (1989).

Self-tuning controllers are a more general class of controllers than MRAC. The control parameters are updated indirectly or directly, based on estimates from an adaptive predictor. In a deterministic environment, the control algorithms which result from a model-following STC and MRAC are very similar (Landau 1981). It must be emphasized that the PAA is the core of all adaptive controllers as it updates the parameters of the adjustable controller in direct schemes and updates the plant model parameter estimates in indirect schemes.

Despite the significant developments that have been made and the impressive applications to several practical problems that have been demonstrated, it must be noted that a characteristic of the existing adaptive control procedures is their model dependence, i.e., the requirement for explicit *a priori* specified model structures. Consequently, although convincing and well-developed algorithms exist at present for linear systems, only very preliminary results are available for more general (for instance, nonlinear) systems. One of the principal directions in which research in the adaptive control of nonlinear systems is progressing is in the employment of artificial neural networks by exploiting their capability for approximating the input-output mappings of dynamical systems. The possibilities offered for the identification of even highly complex dynamics without explicit model dependence have made neural network-based schemes attractive alternatives for the development of adaptive controllers. Furthermore, the parallel computation features of neural networks often translate into speed advantages in the

F:7.2-3 a,b

General Adaptive Control Structure

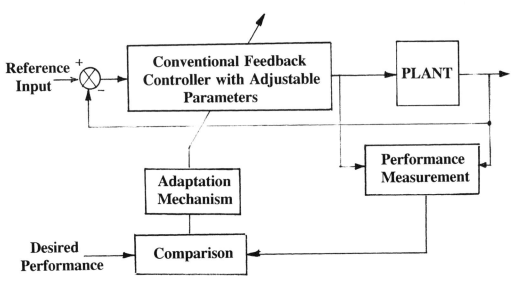

Model Reference Adaptive Control Scheme

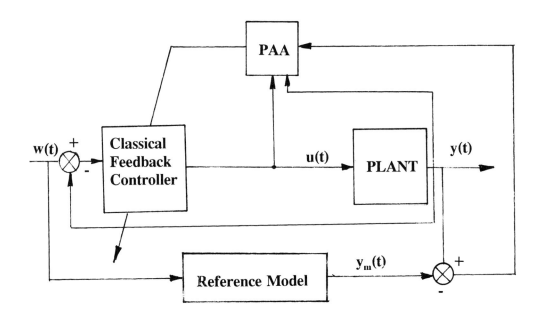

identification and control computation at each step of implementation, when compared to traditional adaptive control algorithms (Sudharsanan and Sundareshan, 1991; Karakasoglu et.al. 1991; Muhsin et.al. 1992).

7.2.2.3 Distributed (or Decentralized) Control Systems

The complexities in the design of control systems get considerably reduced if all the information about the process being controlled and the calculations based on this information are "centralized", i.e., take place at a single location. For a precise understanding of the role of the information in the computation of control policies, it is generally useful to distinguish two kinds of information that are available:

(i)   Information about the System Model -- such as coefficients of a "transfer function". Since this remains constant for any system, we may call this *off-line* or *a priori* information.

(ii)   Sensor information about the system response -- measurements of outputs or states. This is *on-line* information in a dynamical system.

When considering the control of complex large-scale systems, the presupposition of centrality fails to hold due either to lack of centralized information or lack of centralized computing capability. It is very simple to see that although the model information itself may be centralized, requiring the on-line information at a central location is not good from economic reasons. An example is a freeway traffic system where miles and miles of communication channels are needed to transmit the data collected at various sections to the central computer.

Thus for economic and sometimes political reasons, we cannot have a centralized information structure or in other words, we need to live with "information constraints" which make bits and pieces of information available at different locations in the system making "centralized approaches" useless.

The problems arising from the presence of information constraints can be overcome by a process of *decentralization* where the information processing and computation is distributed over the locations where information is available. Thus, referring to Fig. 7.2-4a, if the states x_1, x_2 and x_3 of three parts of the overall system can be measured at locations L_1, L_2 and L_3, then we attempt to perform the computations at these locations and implement the decentralized controllers C_1, C_2 and C_3 at these places. The availability of efficient and inexpensive microprocessors in the present times makes such distributed control strategies more and more attractive.

Another principal reason for choosing to implement a distributed control strategy of the form shown in Fig. 4a is the high degree of local autonomy provided to the different parts of the overall process being controlled. It must be noted that an important characteristic of intelligent control (which will be discussed in a greater detail later) is the local autonomy enjoyed by the subsystems constituting the overall process being controlled.

Depending on the extent of information exchange permitted between decentralized controllers, *one has different degrees of decentralization* ranging form complete decentralization to complete centralization.

(Complete   centralization   ⟶   Everyone   possesses   all information

Complete   decentralization   ⟶   Information   sets nonoverlapping).

Quite simply, if a large system is to be controlled by two controllers C_1 and C_2 and the total state x is divided into two parts x_1 and x_2, and if x_1 is not available to C_2 and x_2 is not available to C_1, then by necessity we should design controllers in the form

$$u_1 = f_1(x_1)$$

and

$$u_2 = f_2(x_2)$$

which are completely decentralized. Alternately, one might require some information exchange between C_1 and C_2 to improve the performance.

Occasionally in many physical systems such as industrial processes and chemical refineries, decentralization is *required due to the delays in information transfer* (when the information may lose its utility). Some advantages resulting from decentralization are enhanced reliability, maintainability and modularity of the system components (Sundareshan, 1977b).

F:7.2-4 a,b

Distributed Control Strategy
Decentralized
Controllers

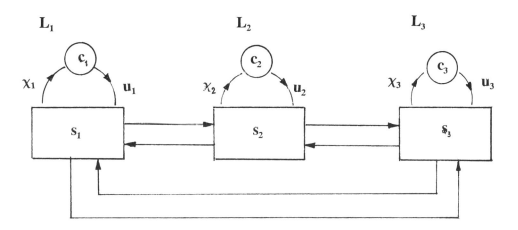

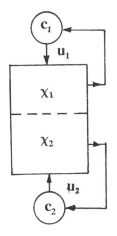

Single Process Controlled by Two Controllers

The employment of a decentralized decision-making strategy as opposed to a single centralized decision-making unit raises several complex issues, the principal one being the establishment of an efficient coordination mechanism between the several decentralized decision-makers. The inefficient operation of several large-scale systems can be attributed to lack of fundamental understanding and modeling of the underlying interactions and the lack of coordinated decision-making strategies. To briefly illustrate the need for such coordination, consider a typical urban traffic grid of one-way streets, where the timing and duration of the green, red and yellow cycles for the signal light at each intersection are computed locally based on the measurements of the queue lengths in the two local one-way links. In such a system, it is evident that a good coordination of signals is necessary in order to achieve the objective of smooth traffic flow, without unnecessary waste of time and energy consumption. For instance, one would like to coordinate adjacent signals by dynamically synchronizing them through appropriate communication of queue lengths so as to minimize the conditions of stop and go. The need for such coordination exists in a very similar manner in several other queuing systems ranging from a single batch-processing computer system to vast interconnections of computer and communication networks and ranging from a single programmable machine on the shop floor to a vast flexible manufacturing system.

Explicit concern with problems of coordination and decentralized decision-making and their relationship to organizational structure and effectiveness has led to the advent of hierarchical system concepts. Ever since the pioneering work of Mesarovic et.al. (1970), which represents the first formal attempt at characterizing the structural properties of hierarchical systems, there have been a very large number of reports devoted to different types of studies on these systems, too vast to reference here. Some representative works that highlight the design of decentralized controllers for different types of systems with diverse control objectives are available in (Sundareshan, 1977a and 1977c; Ho and Mitter, 1976; Findeisen et.al., 1980).

7.2.2.4 Optimal Control Systems

Optimal controller designs are needed when the control task must be executed in an "optimal" manner, i.e., the desired system response must be realized while optimizing a specified performance measure. A general problem formulation is as follows.

Optimal Control Problem: Given the dynamical description of the process being controlled as $\partial x/\partial t = f(x(t), u(t), t)$ where $x(t)$ represents the state of the system at instant t and $u(t)$ is the control signal to be determined, the problem is to find the control $u^{*}(t)$ that minimizes a chosen performance measure

$$J(x(t), u(t)) = \int_{t_0}^{t_f} g(x, u, t)dt,$$ where t_0 and t_f denote the staring time and the final

time relatively. Functions f and g are known nonlinear functions obtained from modelling the process dynamics and the performance desired.

The problem stated above is very general and includes several problems of practical relevance as special cases. For instance, the "minimum time optimal control problem," where the objective is to make a transfer of the system state from the known initial value $x(t_0) = x_0$ to a specified final value $x(t_f) = x_f$ can be obtained by selecting $g = 1$, which results in the performance measure to be minimized $J = (t_f - t_0)$. Similarly, for the "minimum control energy problem," where the objective is to make the transfer from the initial state x_0 to the final x_f with the minimum expense of control energy, function g is selected as a measure of the control signal, eg. $g = u^2(t)$. In this case, the performance measure J reduces to the energy contained in the control signal $u(t)$.

A considerable literature exists on the optimal control of different types of dynamical processes with various performance measures and several texts are available on this topic (eg. Sage and White, 1978 and Stengel, 1986). Solution of optimal control problems generally require sophisticated mathematical tools such as Dynamic Programming (Bertsekas, 1987) and Calculus of Variations (Gelfand and Fomin, 1963). These solutions are typically of an iterative nature and require computer-based procedures. For special classes of problems, such as when the system being controlled is a linear one and the performance measure to be minimized is a quadratic function, explicit analytical solutions resulting in state feedback control functions are possible. The problem of designing optimal controllers become considerably more complex if additional requirements are included, such as adaptivity to parameter variations and distributed control implementation.

7.2.3 Intelligent Systems

7.2.3.1 What Constitutes Intelligence in Physical Systems?

The desire to satisfactorily analyze and design engineering systems of ever increasing complexity and the demands for increased performance and versatility which push system requirements beyond the capabilities of conventional adaptive methods have given rise to the concept of Intelligent Systems. Most critical of the requirements in these systems is the demand that the system performs autonomously in highly uncertain environments.   To deliver the desired performance, these systems must be able to learn about their own inherent capabilities, learn about their environment and about their inputs and be able to reconfigure themselves according to the information they have gained.

In these systems, intelligence commonly refers to the specific processing the system executes.   This includes various kinds of algorithms for data processing, analysis and decision-making (including control).   The nature of these algorithms could be quite diverse ranging from numerical processing (as for signal processing, control or optimization objectives) to symbolic and qualitative processing (such as those being executed in an expert system shell). Specific architecture issues could attain significant relevance in this context, a typical illustration being the processing that involves sensor data fusion.

Building intelligence into a system is often done in practice by collecting human expertise and coding it in (as rules or logic instructions).   While satisfactory designs of intelligent systems require precise mathematical modeling and organized use of established mathematical techniques of control and signal processing, in current industrial practice a popular approach to building intelligence is in the form of adding application software which may consist of some scheduling and planning algorithms, exception handling procedures for abnormal operating modes and safety processing.   A major portion of this effort often involves heuristic rules that are proposed by human operators.   Contemporary research, however, is placing emphasis on the utilization of precise mathematical tools and principles in the realization of intelligent system objectives.

7.2.3.2 Attributes and Goals of Intelligent Control

For engineering applications, incorporation of intelligence is no more crucial than in the design of intelligent devices and controllers. The principle of feedback to control dynamic systems is fundamental and has given rise to specific procedures such as optimal control, adaptive control, etc. to meet specific control objectives. As these methods have found their way into standard practice, they have permitted consideration of a wider spectrum of complex applications. These are characterized by uncertain models, high degrees of nonlinearity, distributed sensors and actuators, high noise levels, abrupt changes in dynamics, hierarchical and distributed decision-makers, multiple time scales, complex information patterns, large amounts of data and stringent performance requirements. The degree to which a control system deals successfully with these complexities depends on the level of intelligence in the system.

Hence the attributes of an intelligent control system (Soureshi and Wormley, 1990) could be briefly listed as:

(i)	High degree of autonomy;
(ii)	Reasoning under uncertainty;
(iii)	Performance in a goal-seeking manner;
(iv)	Higher level of abstraction in the application of feedback;
(v)	Data fusion from a number of sensors;
(vi)	Learning and adaptation in a heterogeneous environment.

Also, the commonly stated goals of intelligent controllers are:

(i)	Autonomous response to discrete symbolic commands;
(ii)	On-line identification of changes in system structure from processing sensor data;
(iii)	Facilitating use of heterogeneous knowledge sources (qualitative, symbolic, etc.) and multiple sensors;
(iv)	Self-adjustment and reconfiguration due to changes in system structures.

7.2.3.3 Intelligent Control Based on Neural Nets

Although the requirements for an intelligent controller are quite extensive, as Werbos (Werbos, 1992) aptly points out, the mammalian brain is a living proof that one can use analog distributed hardware to build a controller that possesses a number of attributes stated above. In particular, he emphasizes that the brain as a whole system can be viewed as a neurocontroller that combines many remarkable properties which include the following:

(i)   The ability to plan, optimize or accomplish goals over long-time horizons;
(ii)   The ability to learn in real time;
(iii)   The ability to control millions of muscles or motors in parallel;
(iv)   The ability to adapt to a highly nonlinear environment;
(v)   The ability to cope with noise.

Contemporary research on neural networks, both biological and artificial, can hence find crucial applications in building controllers with "brain-like capabilities." In particular, studies on the biological system which focus on characterizing and validating the information flows in the cytoskeleton could provide a fundamental understanding of the mechanisms of learning and intelligence. Although the mainstream efforts in neuroscience and neural modeling have relied on the classical neuron model which attempts to characterize the knowledge only in terms of voltages in cell membranes and inter-cellular synapse strengths, some recent works (Conrad, 1984; Hameroff, 1987) that have employed the cytoskeletal information processing capabilities could provide very valuable information in this direction. Particular reference should be made to the studies on the role of microtubules (MTs) and microtubule associated proteins (MAPs) in information processing and the dynamical interactions (in the context of molecular automata) of coherent phonons in MT/MAP networks, cooperative dynamical functions resulting from the arrangement of tubulin dimer subunits in appropriate lattice symmetry (Koruga, 1984) and the intrinsic electric field effects in MTs, its interaction with dipole oscillations in dimers and the control effects of externally applied electric fields on the information carrying properties of MTs.

In a similar manner, a number of specific studies undertaken in the realm of artificial neural networks, and particularly their application to neurocontrol, can be employed for realizing the objectives of intelligent control. One of the important requirements of intelligent control is to utilize the past behavioral information in order to predict the control to be applied and this can be handled by a neural network used in the inverse control mode. Also, it is widely known that a key problem in backpropagation training of a neural net is that the training has to be in the off-line mode and hence true real-time learning becomes difficult. Some of the adaptive critic designs (Barto et al, 1983; Werbos, 1991) attempt to overcome this problem at the cost of exactness and simplicity of backpropagation. It is becoming increasingly realized that hierarchical decision-making structures are to be employed for realizing the full potential of intelligent control and that neural networks are more useful at the lower levels of this structure where representation of system dynamics is of importance.

A question of particular importance in tailoring an intelligent control architecture is how to delegate the appropriate functions to the component units of the control architecture by assigning appropriate processing technology based on (i) the availability of knowledge for processing, and (ii) the objectives of control. Much of the currently popular work in this direction is in combining the two specific technologies - neurocontrol and fuzzy logic (Kosko, 1992). However, several of the other equally promising approaches such as event-based control, inductive reasoning and artificial intelligence are being suggested as candidate technologies to be included in order to exploit the advantages offered by these approaches in the design of intelligent control mechanisms for specific applications.

7.2.3.4 Knowledge Representation and Processing for Intelligent Control

Central to the realization of the objectives of intelligent control is an organized mechanism for knowledge representation and processing. Such a mechanism should include appropriate transformations between diverse knowledge types, such as numeric, symbolic, linguistic, etc., residing at different parts of the generic intelligent control architecture in order to provide proper interfacing in the execution of the overall control task. While traditional control design procedures have paid a greater degree of attention to purely numerical knowledge processing, relatively little effort has been given to the processing of other forms of knowledge.

Special characteristics of the specific knowledge type required to be processed at a given stage should be given proper attention in tailoring the processing characteristics at that stage. It is evident that for realizing an automated implementation of intelligent control, processing of symbolic data is highly input-output intensive and usually requires more memory references and input-output operations than numerical computations. Equally evident is that there are some spects in the control design process which are not naturally amenable for representation by precise and numerical methods, but can be more efficiently represented by heuristics or rules of thumb.

A general functional hierarchy in the overall processing of knowledge for the control of a complex dynamical system is shown in Fig. 7.2-5. This figure also identifies the principal features of the represented knowledge at the various stages and the possible tools for processing of this knowledge.

7.2.3.5 A Hierarchical Framework for Decision-Making in Intelligent Control

The primary goal in the design of an intelligent control structure that attempts to mimic the cognition and task execution performance by humans is the inclusion of the capabilities for fault diagnosis and reconfiguration, effective planning (long-term optimization) under conditions of noise and qualitative uncertainty, simultaneous control of a large number of variables and data fusion. Furthermore, in order to achieve autonomous operation, the controller must have the capability of performing a number of functions in addition to the conventional functions of regulation and tracking. A hierarchical structure that can effectively perform task decomposition and distribute the overall decision-making process into decision-making at several levels can provide a framework for realizing the above mentioned capabilities. Such a structure can be tailored by employing the principles and tools from hierarchical and multilevel system theory (such as the principle of increasing intelligence with decreasing precision, goal and model coordination, etc.).

A generic architecture for intelligent control that employs a hierarchical structure is shown in Fig. 7.2-6. This architecture can be regarded as a generalization of some well-known structures such as the action-critic control structure and the master-slave formulations that are popular in neural network-based processing. In the structure shown in Fig. 7.2-6, Level U represents the processing of higher level functions while Level L encompasses the remainder of the functions leading to the exact determination of the control variables to be input into the system to realize the desired objectives.

It must be emphasized that each of the two levels shown could be decomposed into a number of sublevels appropriately for a specific application.

Some characteristics that guide in the determination of the number of sublevels are the types of knowledge available for processing, the control objectives, the complexity of mathematical representation of system dynamics and the extent to which human operators can intervene in the system operation.

An organized and systematic representation of the performance at the individual levels can be obtained by specifying appropriate models (such as fuzzy maps, queuing models, Markov chains, etc.) for knowledge processing at these levels and establishing a correspondence with the functional hierarchy for knowledge representation and processing shown in Fig. 7.2-5.

7.2.4 Neuromolecular Computing and Information Processing

Recent efforts at explaining the human cognition process through connectionist models of functional brain organization have provided strong links between neuroscience and computing and information processing technology. Popularly used neural network models, however, are overly primitive and assume neurons and their synaptic connections to be the fundamental units of information processing, with the neuron functions modeled by switches and the neural net learning approximated by modifications to the synaptic weights. Contrary to this populistic view, some organized research efforts are presently under way to utilize the complex internal structure of neurons and synapses in order to approach the mechanisms of brain cognition by consuming the dynamic regulatory activities within individual neurons and the underlying molecular level information processing via cytoskeletal functions.   These studies (Hameroff, 1987; Conrad, 1985; Rasmussen et al, 1990; Koruga, 1992) have introduced the idea of molecular automata within neurons as a basis for intracellular information processing and have proposed models of cytoskeletal automata subserving neural network in the collective neuronal functions ultimately relating to cognition.

A significant portion of the past and present work in the theoretical analysis of brain functions and neural networks has been motivated by the desire to build better parallel processing machines than what are possible via von Neumann technology.   This work has progressed following both mathematical and nonmathematical (studies in biology and psychology) directions. The mathematical studies have employed various types of models such as algebraic (for example, McCulloch-Pitts neuron model), calculus-based (for example, Hodgkin-Huxley equations) and geometrical.   While the geometry-based studies that have employed Euclidean geometry (such as vector analysis and differential geometry of cartesian vectors) have yielded satisfactory design principles for synthesizing von Neumann parallel computers, it is in the utilization of non-Euclidean geometrical principles (metrical, fractal and chaotic) lies the full power of realizing the capabilities of non-von Neumann neurocomputers.

Innovations in technology by drawing parallels from studies on the biological system have been progressing very rapidly especially during the past five years. Various implementations (optical and electrical) of artificial neural networks have been realized for parallel computing machines. Very active research and development programs in VLSI implementations are presently under way leading to the advent of various types of neuroboards and neurochips (Hecht-Nielsen, 1990; Maren, 1990). Nevertheless, the current approach to neurocomputing is a very simplified one, which follows the basic principle that there are multiple entries into the neuron and there exists only one exit (off or on) from the neuron. The parallel processing capabilities come from a multilayer network arrangement of such neurons with one input layer, one output layer and several hidden layers. Although such network architectures and their implementations have yielded satisfactory performance in specific applications, it is becoming increasingly realized that the inclusion of the dynamical changes within the individual neurons would provide better mechanisms for neural computation. In particular, a focus on the molecular computing within the neuron (effect of the cytoplasm on changes occurring in the nucleus) and the molecular information processing occurring in the cytoskeleton should provide improved structures and intelligent hardware for neuromolecular computing and information processing machines.

The feasibility of such molecular computers is becoming more and more justifiable due to the advent of novel enabling materials (such as carbon-based photonics). These devices have functionalities similar to those of artificial neural networks with the exception that they include molecular processing units which are smaller, faster and possess enhanced overall computational abilities due to the wireless communication links among units by means of specific or nonspecific molecular messengers (Szu and Tate, unpublished). Particularly useful directions of investigation towards the goal of designing and fabricating such devices (molecular chips) are to determine what principles can be extended from the presently existing data on artificial neural networks (Szu and Tate, unpublished) and what types of structural similarities exist between biological material (such as microtubules) and material that could be fabricated (such as

F:7.2-5

Functional Hierarchy for Knowledge Representation and Processing

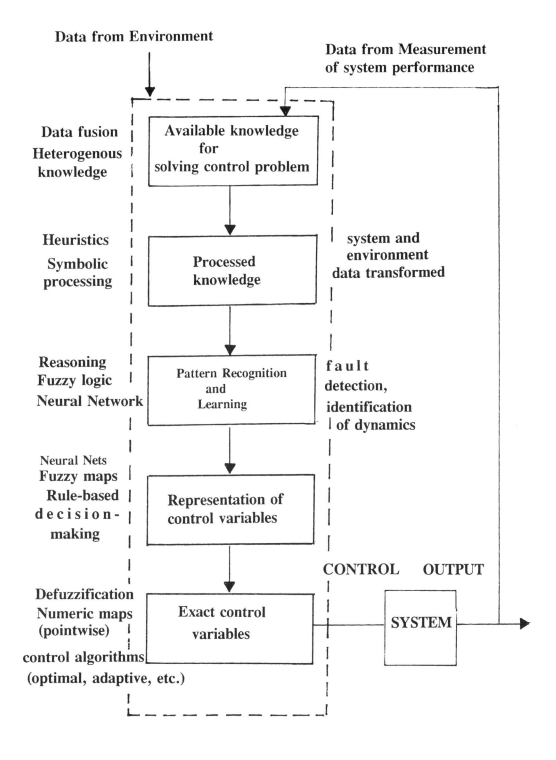

F:7.2-6

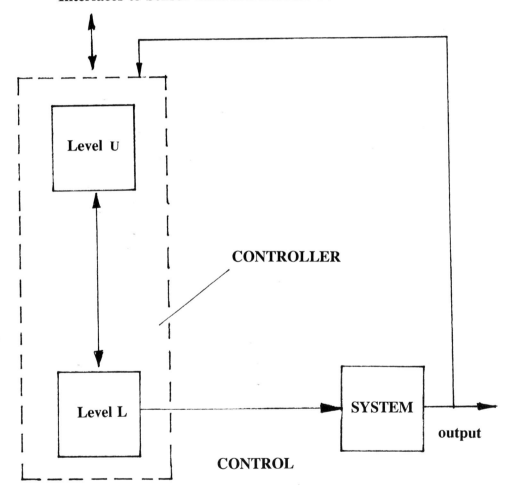

Interfaces to Sensor Data and Human Command

Hierarchical Framework for Decision-Making

Fullerenes (C-60 and other)). Recognition of the fact that molecular computers are non-silicon, carbon-based devices and hence provide a novel implementation of artificial neural networks enables one to synergistically exploit the similarities and differences between these two technologies. Furthermore, the experimentally observed similarities in architecture of biomolecular structures (such as microtubules and clathrin on one hand and Fullerenes on the other hand) could be utilized to design appropriate interfaces and intelligent systems based on nanotechnology.

7.2.4.1 Neuromolecular Approach to Intelligent Control

As mentioned earlier, the information processing models of microtubules by such mechanisms as nanosecond or faster cooperative conformation dynamics for tubulin (*e.g.*, in a cellular automata context), subunit packing symmetry, etc. can help identify more realistic models for the processing elements of artificial neural networks (rather than simple on/off units) which will in turn bring these networks closer to the biological neural system. Furthermore, an evaluation of the real time information processing functions occurring in microtubules to coordinate several complex processes such as mitosis, growth, cell movement, intracellular transport, differentiation, synapse formation and regulation, and other dynamic activities could be utilized in tailoring more efficient learning schemes than are currently popular in artificial neural network literature. These studies together with organized procedures for synthesizing interfaces and molecular electronic devices with the help of the scanning tunneling microscope and nanotechnology should permit a more efficient realization of the intelligent control objectives outlined in Section 7.2.3.

7.2.5 New Computational Paradigms

Vigorous on-going research in the development of novel computational paradigms, which are capable of permitting faster computation together with smaller required memory than the present-day computers, has resulted in several interesting ideas having roots in modern physics, biology and mathematical system theory. In particular, the potential contributions of quantum effects and chaotic systems to computing - particularly to neurocomputing, which can exploit the massive parallelism allowed by such effects are being unravelled. In the efforts to provide an understanding of the computational features underlying real biological neural networks, suggestions have recently been made (Penrose, 1989) that the functionalities of these networks are based on certain fundamental properties of quantum theory. Furthermore, it is argued that these properties lead to computational possibilities at the systems level (as distinct from the device level computations) which are significantly different and more powerful than what are possible with the present day conventional computers (Werbos, 1992). Similarly, connections between massively parallel computing and chaos theory are becoming increasingly clear and many people have proposed that ideas from chaos theory can be used to improve the algorithms of neurocomputing or to provide new devices for implementing them. Furthermore, recent advances in neural networks and nonlinear system theory indicate that new types of computers could be built which do not resemble the classical von Neumann architecture, but which resemble Turing equivalence.

7.2.5.1 Quantum Computing

The desire for developing faster and smaller computers has motivated the research in utilizing quantization and quantum phenomena, such as tunneling, in the realization of computational feats. Quantum theory is already important in the design of microelectronic components. Similar principles offer feasibility in the design of quantum computers which can perform computational tasks that are far superior to what are possible by classical means. Specific quantum phenomena of particular significance in developing new forms of computation are complementarity, nonseparability and quantum interference (Deutsch, 1989). Perhaps one of the first realizations of a truly quantum computing device is that developed by Bennett et al (1992) recently to implement a new form of public key encryption system for secure communication, which could not be implemented by any existing computer.

In order to give a comparison with the capabilities of present day computers, as Deutsch (Deutsch, 1992) points out, quantum computers will achieve their superiority in tasks that do not require the computation of functions (a "function" here denotes an operation or mapping that produces a unique output for each input considered). It is to be noted that several

F:7.2-7

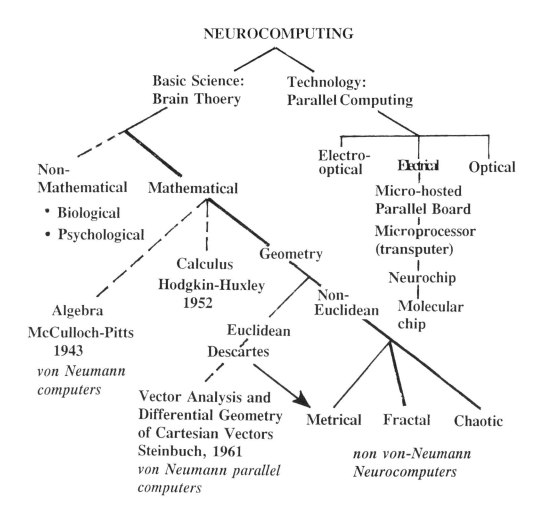

Neurocomputing as a new scientific and engineering paradigm. The ultimate goal is to develop and implement the concept of non-von Neumann neurocomputing in Fullerene C_{60} molecular devices.

computational tasks do not intrinsically require the computation of functions (in the above sense) and rather require solving a problem in the sense of finding a feasible output (that satisfies a specified criterion) for each applied input. Since this output is nonunique, in general, it is not necessary to compute any function and a quantum computer program could output a randomly chosen correct answer much sooner than any present day computer program could guarantee to find an answer.

7.2.5.2 Chaos Computing

Systems, whether arising from the physical arena or the biological arena, generally exhibit static (i.e. spatial) and dynamic (i.e. temporal) properties. The spatial structure can be regular (as arising from homogeneous systems, symmetric systems etc.) or irregular (which are characterized as fractals). Similarly, the temporal behavior could be regular, as in the case of equilibrium (stationary) points and periodic motions, or irregular which is characterized as chaotic motion. The combined occurrence of both types of irregularities - spatial irregularity (fractals) and temporal irregularity (chaos) leads to turbulent states.

The vast research that has been conducted recently on the topic of neural network dynamics has revealed that in the modcling of systems capable of neural computation, three different classes of dynamical systems are of importance (Hirsch, 1989). These are, (i) systems with *convergent* dynamics, where every trajectory converges asymptotically to some equilibrium state, (ii) systems with oscillatory dynamics, where every trajectory asymptotically enters a *periodic* orbit, and (iii) systems with *chaotic* dynamics, where a majority of trajectories do not tend to periodic orbits or converge to a fixed point. It is particularly interesting that although biological systems practically never display convergent dynamics, the most popular neural computation models are synthesized to be convergent due perhaps to the exceeding difficulty in studying the dynamical properties in the other cases.

Mathematical models that lend themselves appropriate for studies on chaos computing are nonlinear differential equations in the state vector $s(t)$ of the dynamical system expressed in the form

$$\frac{\partial s}{\partial t} = f(s(t), \ u(t), \ w(t)) \qquad\qquad (7.2.5)$$

where u(t) denotes the vector of input excitations and w(t) denotes the vector of system parameters (which include the synaptic connection weights in the case of a neural network). For neutral computation, the problem of particular interest in these systems is the pursuit of Turing-equivalent processing i.e. construction of Turing machines from systems characterized by the dynamics specified by Equation 7.2.5. A point of significant difference from the conventional computers which are binary state machines is the continuous state properties of these machines.

One particular line of study that has attracted considerable attention in the recent past is the design of systems (Baird and Eeckman, 1992) that possess dynamic attractors which provide superior feats of computation. It is vigorously pointed out in (Baird and Eeckman, 1992) that recurrently interconnected associative memory modules provide an architecture which can store and recall multiple oscillatory and chaotic attractors and hence provide a feasible framework in which to arrange and exploit the special capabilities of dynamic attractors.

7.2.6 Optimal Control at the Molecular Scale

There exists considerable literature on the use of tools from control theory, in particular optimal control, for actively controlling the dynamics at the cellular level. A majority of these studies are directed to the analysis of the biological and physiological processes at this level and are motivated by applications to cancer cell kinetics and therapy (Sundareshan and Acharya, 1984; Sundareshan and Fundakowski, 1986), regulation of immune responses (Perelson, 1989; Varela and Coutinho, 1991), etc. However, corresponding efforts at the design of control strategies aimed at material modification at the molecular level are not very extensive and are being proposed only in the recent times. Drawing motivations from applications in the control of chemical processes, and specifically aimed at the control of laser optical fields (Tannor and Rice, 1985; Shapiro and Brumer, 1986; Kasloff et al, 1989; Jakubetz et al, 1990; Park et al, 1991), these studies are providing confirmation that application of techniques from modern control engineering at the level of quantum systems could open up a new domain of technological advances.

7.2.6.1 Basic Characteristics of Molecular Level Control Designs

Design of appropriate strategies for excitation of molecular scale physical systems to achieve prescribed molecular motion is a topic of significant importance. In generic terms, design of control in these systems typically involves an appropriate tailoring of an electric or optical field to achieve selective chemistry, such as to break a particular bond within a polyatomic molecule. In the popular literature on physical chemistry, there exist numerous experimental works where such problems have been attempted by using approaches guided mainly by intuition. For instance, in order to break a specific selected bond within a molecule, one would employ an excitation at the frequency associated with that bond to induce a resonance that would ultimately break that bond. However, actual implementation of such intuition-directed solutions could be far from simple and may lead to several pitfalls. An example of the difficulty faced is the complexity of localizing the energy that needs to be imparted to the molecule within the selected bond and the interactions of the bond with the rest of the molecule.

Control of molecular dynamics by appropriate laser field designs to achieve prescribed molecular motion can lead to several potential technological payoffs. Some specific applications that could benefit significantly from progress in this area are the design of ultrafast computer memories, high density encoding and decoding for information processing, creation of new molecules by molecular scale surgery, etc. (Rabitz, 1989). Efforts at developing systematic approaches for application of sophisticated tools from mathematical control theory in this area are indeed important and some progress in this direction has been made within the last five years.

To be exact, there have been numerous earlier attempts at the design of appropriate excitation strategies to achieve specific objectives in molecular scale physical systems. These, however, have relied mostly on intuition and guesswork (of course drawing motivations from physical considerations). Some specific examples that may be cited are the work directed to the design of pulse strategies for nuclear magnetic resonance (NMR) imaging sequences (Garroway et al, 1974; Sutherland and Hutchinson, 1978; Hoult, 1979) and the efforts at the tailoring of laser fields for selective excitations in polyatomic molecules (Bloembergen and Zewail, 1984). While these studies provide commendable preliminary attempts at molecular scale systems analysis and control, an assessment of the efficiency of these procedures (in terms of optimality of the control schemes) is not available.

From the perspective of control system designs, some work that has been reported more recently on the evaluation of the possibilities for controlling the selectivity of chemical reactions through a careful manipulation of laser fields is of interest. Following an approach that employs a variational formulation, Tannor and Rice (1985, 1987) propose a scheme for achieving selectivity of reactivity by a two-pulse process. In this scheme, the first laser pulse is used to excite the molecule into a nonstationary state which is then allowed to evolve freely. The second pulse is then used to stimulate emission to bring the molecule back to a desired state. A scheme for coherent excitation of a molecule using two or more lasers is proposed by Holme and Hutchinson (1986, 1987). A different approach for designing a pumping field that involves two lasers in order to drive the state of a molecule into degenerate dissociative product states is outlined by Shapiro and Brumer (1986). In a related work, design of an algorithm for rf-pulse design to achieve selective excitation in nuclear magnetic resonance (NMR) is discussed by Connolly et al (1986). Owing to the fact that previous attempts at laser induced selectivity have not generally achieved intended goals due to a rapid intramolecular redistribution of vibrational energy, such approaches that attempt to appropriately utilize the system dynamics in the field design process have attained some degree of popularity.

A polyatomic molecule is a complex quantum mechanical system and for highest efficiency, the designed external field must work cooperatively with the molecular system in attaining the desired objectives. Detailed properties of such quantum systems that can be established using the general formalisms described by Butkovskii and Samoilenko (1979) are hence very useful. In particular, in achieving efficient field designs, one may be guided by the controllability properties and feedback characteristics of quantum dynamical systems studied by Davies (1977) and Huang et al (1983). More recent work by Rabitz and coworkers (Rabitz, 1989; Shi et al, 1988; Pierce et al, 1988; Dahleh et al, 1990) on the optimal control of molecular motion has put this topic on a sound theoretical footing and has demonstrated that the optimal control framework provides a rigorous basis for assessing the practicality of achieving specific molecular objectives and the means to design external fields in specific cases.

As noted earlier, the principal objective for control at the molecular scale is the design of appropriate fields (electric or optical) that have the ability to selectively break molecular bonds or to activate a reagent that can preferentially react at a chosen site. A number of associated constraints need to be taken into account in achieving this objective, however: the desire for minimal disturbance or damage to the remainder of the molecule; minimal expenditure of laser energy, etc. As established in applications to similar problems in other areas (such as aircraft and spacecraft control [Bryson and Ho, 1975]), optimal control methods seem to be particularly suited for handling these tasks. The assessment of molecular level interactions and the control of dynamical processes at this level can, however, pose unique new challenges to traditional control design procedures.

Analytical formulations of the control of quantum dynamical systems lead to complexities in solutions due to the infinite dimensional nature of the system. The evolution of the state in these systems and the dynamical interactions are required to be modelled by partial differential equations to account for the spatial and temporal changes. Although there is a wealth of unified theory for studying the problems of modelling, qualitative properties and optimization of lumped systems, for systems of the distributive type there does not exist such a unified theory due to the mathematical nature of the associated problems and the existence of many different types of qualitative behavior. Nevertheless, analysis and controller design for distributed parameter systems arising in diverse application areas have been investigated for a number of years. To cite some illustrative works, Seinfeld (1970) and Desalu et al (1974) have discussed optimal strategies for air pollution control by using models which are in the form of three-dimensional time-dependent partial differential equations. Ergen (1954) discusses models that arise in nuclear reactor dynamics which take the form of functional differential equations. Theoretical developments specifically addressing problems of particular interest in controlling such systems are available in Butkovskii (1967), Triggiani (1975), Balakrishnan (1971) and Delfour and Mitter (1972). These and other more recent developments available in control theory literature provide tools for handling problems arising in the control of quantum mechanical systems.

Problems of particular complexity in these systems are the development of analytical models and the unavailability of satisfactory identification tools. As noted by Rabitz (1989), such complexities cannot be overcome by raw computing power alone and the use of sensitivity analysis and order reduction procedures are to be given careful consideration. In particular, such questions as how the interactions and parameters at both the intermolecular level and the intramolecular force field level influence the properties at the overall system level need to be addressed. While overly simplified model structures resulting from spectral information are possible, more accurate optimal field designs necessitate inclusion of a number of other issues and effects such as molecular rotation and appropriate assessments are to be made of the sensitivity to molecular Hamiltonian and dipole function uncertainty. The difficulty in modelling these effects would necessitate the design of robust control schemes that provide a sufficient degree of tolerance to the uncertainties and other factors arising from ignorance of dynamical phenomena.

The use of quantum mechanical models of polyatomic molecules for designing molecular control strategies by using traditional control theory approaches leads in general to rapidly increasing computational complexity as the number of atoms in the molecule increases. Furthermore, analytical solutions for the optimal control of systems with a distributive structure are not available and one needs to employ a numerical framework for evaluating the solution.

The optimal control formulation for the quantum molecular control problem leads in general to a multiplicity of solutions (countably infinite). The possibility of alternate solutions, instead of a single unique solution as in the case of the control of lumped linear systems, can lead to attendant complexities of characterizing them. However, this could have some beneficial aspects as well. In particular, one will have the freedom to select specific solutions that offer greater convenience in implementation. For instance, control fields that are more convenient to design can be picked out.

A special characteristic of the quantum molecular systems which makes the control problems challenging is the very short time-scale (typically 10^{-9} to 10^{-15} sec) in which the dynamics evolve. Such ultrafast dynamics can pose special problems for control implementations, such as rendering the process of monitoring the evolution of the molecular state and feeding it back for control (i.e. redirecting the field) almost impossible. The required open loop nature of the control hence places a great demand on the soundness of the mathematical models and the preciseness of the control algorithm. The fast evolution of the dynamics can offer some benefits in conducting specific identification studies. For illustration, in the study of a laser-stimulated chemical reaction, one can exploit the high speed nature of the evolving dynamics in obtaining a larger amount of learning data within a short period of time, and consequently in tailoring experiments aimed at identifying specific quantities such as the Hamiltonian of the system.

Notwithstanding the complexities and challenges cited above, some progress is being achieved in the very recent times for handling field design problems by employing variational techniques from optimal control theory.

Development of numerical techniques for solving the dynamical equations and the optimization problems has received attention and particular focus has been given to the solution of the time-dependent Schrödinger equation and the corresponding optimal control problem. Pioneering studies are reported by Rabitz and co-workers (Shi et al, 1988; Pierce et al, 1988; Dahleh et al, 1990), some of which will be briefly outlined in the following sections for illustrative purposes. For future work in this area, one may note that in the control of dynamical systems with complex (for instance, nonlinear) dynamics, a mathematical tool that has resulted in significant payoffs is differential geometry. Possible reformulations of the molecular control problem using geometric methods could offer simplifications in the search for simple implementable solutions.

7.2.6.2 Computation of Optimal Control for Selective Vibrational Excitation

The principal goal in selective vibrational excitation in a polyatomic system is to achieve a desired bond energy distribution with a minimum disturbance to the remainder of the system. For an analytical formulation of this problem, one can use the fact that the molecular system dynamics is governed by Schrödinger's equation, which facilitates a reformulation of the problem as a constrained optimization problem for which a numerical solution can be developed. In the following, we shall outline an approach described by Shi et al (1988) for the design of an optical field for selective vibrational excitation in linear harmonic chain molecules. For facilitating an easy reference to the original source, we will employ the same notations as far as possible.

Consider a linear harmonic chain molecule consisting of $(N+1)$ atoms. Let m_i denote the mass of the i-th atom and k_i denote the force constant for the i-th bond. Also, let the two N-dimensional vectors $q(t)=[q_1(t)\ q_2(t)\ \ldots\ q_N(t)]^T$ and $p(t)=[p_1(t)\ p_2(t)\ \ldots\ p_N(t)]^T$ where $q_i(t)$ and $p_i(t)$ respectively denote the displacement and the corresponding momentum for the i-th bond. Then the molecular Hamiltonian can be expressed as

$$\mathcal{H}_m = \frac{1}{2}p^TGp + \frac{1}{2}q^TFq \qquad (7.2.6)$$

where G is a symmetric NxN matrix with the (i,j)-th element (i.e. element in the i-th row and the j-th column) g_{ij} defined by

$$g_{ij} = \delta_{i,j}\left(\frac{1}{m_{i+1}} + \frac{1}{m_i}\right) - \delta_{i,(j+1)} \cdot \frac{1}{m_i} - \delta_{i,(j-1)} \cdot \frac{1}{m_{i+1}} \qquad (7.2.7)$$

and F denotes the NxN diagonal matrix of force constants, $F=\text{diag}[k_1 \ k_2 \ . \ . \ . \ k_N]$.

The control problem of interest is the design of an external optical field which ensures selective excitation of a particular bond with minimal disturbance to the rest of the molecular system. More precisely, the requirements can be specified as follows. For the chosen bond s and a specified final time T,

(i) the expectation value $<q_s(T)>$ attain a specified value Q_s;

(ii) the corresponding expectation value $<p_s(T)>$ is positive; and

(iii) the total energy of the molecular system remain as small as possible during the time interval $0 \le t \le T$.

In order to express the dynamical equations in the form of a linear state-variable model, with a classical description of the external optical field, the total Hamiltonian for the molecule in the control field u(t) with the polarization along the molecular chain can be written as

$$\mathcal{H} = \mathcal{H}_m + \mathcal{H}_{in} = \frac{1}{2}p^T G p + \frac{1}{2}q^T F q - D(q)u(t) \qquad (7.2.8)$$

where D(q) is the molecular dipole function. Let D(q) be approximated by retaining only the first two terms in the expansion as

$$D(q) = D(0) + \Delta \cdot q \qquad (7.2.9)$$

where $\Delta = \partial D/\partial q]$ $q=0$, $\partial D/\partial q$ being the gradient vector $\partial D/\partial q = [\partial D/\partial q_1 \ \partial D/\partial q_2 \ . \ . \ . \ \partial D/\partial q_N]$.

The equations of motion for the quantum expectation values of $<q(t)>$ and $<p(t)>$ can then be expressed in the form of two interconnected N-th order vector differential equations as

$$<\frac{\partial q}{\partial t}> \; = \; G<p(t)>$$

$$<\frac{\partial p}{\partial t}> \; = \; -F<q(t)> \; + \; \Delta^T \cdot u(t).$$

(7.2.10)

For a more standard representation, the dynamical equations in (7-24) can be rewritten by introducing the 2N-dimensional state vector

$$x(t) \; = \; \begin{matrix} <q(t)> \\ <p(t)> \end{matrix}$$

(7.2.11)

as

$$\partial x/\partial t \; = \; Ax(t) \; + \; Bu(t)$$

where A is a 2Nx2N matrix and B is a 2Nx1 matrix. We can assume that initially the molecule is in its ground state, providing the initial state value $x(0)=0$ for the dynamical system described by (7.2.11).

For obtaining a measure for the energy of the molecule, the energy of the i-th bond can be defined as

$$e_i(t) \; = \; \frac{1}{2}k_i(<q_i(t)>)^2 \; + \; \frac{1}{2}g_{ii}(<p_i(t)>)^2.$$

(7.2.12)

The total energy of the molecule can then be approximated by the sum of the energies in the various bonds as

$$E_m(t) \; \approx \; E_b(t) \; = \; \sum_{i=1}^{N} e_i(t) \; = \; \frac{1}{2}x^T(t)M_d x(t)$$

(7.2.13)

where M_d is a 2Nx2N diagonal matrix.

The problem of interest can now be explicitly formulated as a "Terminal Control Problem" which can be stated as follows:

Optimal Control Problem: For the dynamical system described by 7.2.13 starting at the initial state $x(0)=0$, find the optimal control $u^*(t)$ that minimizes the cost

$$J = \Phi(T) + L \qquad (7.2.14)$$

where $\Phi(T)=k_s(<q_s(T)-Q_s>)^2+1/2 \quad g_{ss}(<p_s(T)>)^2 \cdot h(-<p_s(T)>)$,   and

$L = \int_0^T [\sum_{i=1}^{N} W_i e_i^2(t) + \frac{1}{2} W_c u^2(t)]dt$ with W_c and W_i, i $= 1, 2 \dots$ N being appropriately specified positive constants and $h(\cdot)$ denoting the unit step function (heaviside function),

$$h(\phi) = 1, \phi>0$$
$$0, \phi<0.$$

The two parts of the cost functional J deserve some explanation. The "terminal cost" $\Phi(T)$ reflects our desire to minimize the deviation of $<q_s(T)>$ from the specified value to be attained Q_s and to minimize the possibility of $<p_s(T)>$ attaining a negative value. The "integral cost" L, on the other hand, reflects our desire to minimize the weighted bond energy and the control energy (pumping field) supplied from outside. The selection of the weighting constants can be appropriately made to vary the relative contributions of the various terms in the overall cost.

The problem stated above is a standard linear-quadratic (LQ) optimal control problem and hence a solution can be readily obtained by using variational calculus methods (Sage and White, 1977). Since the problem is a constrained minimization problem, where the minimization of the cost J is required to be performed under the differential equation constraint (b), the standard procedure is to convert the problem into an unconstrained minimization problem by the introduction of Lagrange multipliers $\lambda_i(t)$, i=1, 2, ..., N and minimizing the "augmented cost"

$$\tilde{J} = J + \int_0^T \lambda^T(t)[Ax(t) + Bu(t) - \frac{\partial x}{\partial t}]dt, \qquad (7.2.15)$$

where $\lambda^T = [\lambda_1(t) \ \lambda_2(t) \ ... \ \lambda_{2N}(t)]$. It is evident that when the constraint equations (b) are satisfied, the control $u^*(t)$ that minimizes J also minimizes J.

Following the well-known standard solution procedure (Sage and White, 1977), the necessary conditions to be satisfied by the optimal control can be written down as,

$$(i) \quad \frac{\partial \tilde{J}}{\partial u(t)} = w_c u(t) + \lambda^T(t)B = 0 \qquad (7.2.16a)$$

$$(ii) \quad \frac{\partial \lambda}{\partial t} = -A^T \lambda(t) - W_d M_D x(t) \qquad (7.2.16b)$$

$$(iii) \quad \frac{\partial x}{\partial t} = Ax(t) + Bu(t) \qquad (7.2.16c)$$

and

$$(iv) \quad \lambda_i(T) = \begin{array}{l} k_s(<q_s(T)-Q_s>) + g_{ss}(<p_s(T)>)^2 \cdot h(-p_s(T)) \qquad (7.2.16d) \\ 0, \ i \neq s. \qquad\qquad\qquad\qquad if \ i=s \end{array}$$

The matrix W_d in the "costate equation" (7.2.16b) is defined by $W_d=diag[w_1 \ w_2 \ ... \ w_N]$.

For computing the optimal control $u^*(t)$, the system of equations specified by 7.2.16a - 7.2.16d) are to be solved. Observe that 7.2.16a provides an expression for $u^*(t)$ in terms of the "costate vector" $\lambda(t)$, which can be evaluated by solving the two 2N-dimensional coupled equations 7.2.16b and 7.2.16c. This is, however, a two-point boundary value problem since the initial condition for $x(t)$ is specified as $x(0)=0$, while the final condition for $x(t)$, given by 7.2.16d, is to be used. Hence an analytical solution cannot be obtained and one thus resorts to a numerical procedure for computing the optimal control $u^*(t)$.

A number of stable numerical algorithms are available for the solution of such two point boundary value problems (Luenberger, 1969; Zangwill, 1969). In Shi et al (1988), the conjugate gradient algorithm is used for the computation of $u^*(t)$ and several characteristics of this solution are studied. A number of specific examples of linear harmonic chains with different numbers of atoms are also analyzed for illustrating how the optimal design of the optical field for achieving selective vibrational excitation in polyatomic molecules can be conducted. Various bond energy profiles are also displayed to demonstrate that local excitation at the specified bond is indeed successfully achieved by the optimal field design. The designed optimal fields in general will have fairly complex shapes. These cannot be tailored by intuition (examining the spectral information, for instance) or guesswork. The complexity of the resulting optimal fields, which are typically very short and intensive pulses, can pose some implementational difficulties, however. Consequently, an evaluation of the sensitivity of the bond energy to errors in the applied optical field is of importance. In Shi et al (1988), such studies are also conducted to demonstrate that any deviations from the desired goals caused by errors in the optical fields are independent of the fields themselves, at least in the case of harmonic systems.

7.2.6.3 Optimal Control of the Schrödinger Equation

While the previous section dealt with the case of designing numerically the optimal control at the molecular level employing linear classical models, a corresponding development can be obtained using quantum mechanical molecular models as well. Unlike the formulation of the problem of controlling the dynamics modelled by an ordinary differential equation as obtained earlier, the distributive structure of the dynamical equation governing the evolution of the state in this case adds to the mathematical complexity and requires a functional analytic framework in which a solution needs to be developed. In this section, we shall briefly outline the solution procedure following the development given in Pierce et al (1988). Since we shall only focus on the formulation of the optimal control problem and its solution, while omitting discussion on the properties of the optimal fields generated, we shall closely follow the notations used in Pierce et al (1988) for easy reference.

Let χ denote the spatial variable taking values in the domain $\Omega \subset R^n$ and let t denote the time variable taking values in the interval $[0,T]$. Let $x(\chi,t)$ denote the state of the quantum system to be controlled by application of the external force field, represented by the operator U which is defined by

$$U\psi(\chi,t) = \int u(\chi,\chi^1,t) \ \chi(\chi^1,t)d\chi^1. \qquad (7.2.17)$$

In 7.2.17, the function $u(\chi,\chi^1,t)$ takes values in a Hilbert space $L_2(\Omega x\Omega)$ for each t, so that U is a Hilbert-Schmidt operator.

Given a target state value χ to be attained at the final time T, the problem of interest is to make the transfer of the state to this value while minimizing the energy due to the external field. Evidently, such a formulation of an optimal control problem includes as special cases several specific problems of interest in applications, such as determining the most efficient laser field for subjecting a polyatomic molecule to achieve transfer of energy to specific sites or to break certain selected bonds.

The evolution of the state in this case is governed by the Schrödinger equation

$$i\hbar\frac{\partial\psi}{\partial t} = (H_0+U)\chi \quad \text{for all } \chi\epsilon\Omega \text{ and } t\epsilon[0,T] \qquad (7.2.18)$$

with the initial condition $\psi(\chi,0)=\psi_0(\chi)$ for all $\chi\epsilon\Omega$. In 7.2.18, the operator H_0 defined by

$$H_0 = -(\frac{\hbar^2}{2m})\nabla^2 + V_0 \qquad (7.2.19)$$

specifies the Hamiltonian of the quantum system being controlled. For specifying an appropriate cost function through inner product operations, let us introduce the Hilbert spaces $X=L_2(\Omega)$ and $X_t=L_2(\Omega x\Omega, [0,T])$. The optimal control problem can then be posed as the "Terminal Control Problem" stated below:

Optimal Control Problem: For the dynamical system described by (7.2.18) starting at the initial state $\psi(\chi,0)=\psi_0(\chi)$, find the optimal control U^* that minimizes the cost functional

$$J(u) = <\psi(\cdot,T) - \hat\psi(\cdot), \psi(\cdot,T) - \hat\psi(\cdot)>_X + W<u,u>_{X_t} \quad (7.2.20)$$

where W is an appropriately selected positive constant and $<\cdot,\cdot>_X$ and $<\cdot,\cdot>_{X_t}$ denote inner product operations on appropriate Hilbert spaces.

Excepting for the generality of the framework in which the problem is formulated (it may be noted that the quantum system being examined can contain an arbitrary number of particles in three spatial dimensions) and the consequently increased mathematical complexity, the optimal control problem is similar to the corresponding problem discussed in the last section. The procedure for obtaining a solution through employing variational arguments follows in a similar manner through the introduction of a Lagrange multiplier function and taking the Frechet derivatives of the cost functional. As before, a set of necessary conditions can be written down which will facilitate evaluating the optimal field to be applied by a numerical procedure such as the conjugate gradient method or a steepest descent search.

The form of the solution is decidedly more complex than that discussed in Section 7.2.6.2 due to the more general framework in which the control problem is handled. For evaluating the detailed properties of the optimal fields, Pierce et al (1988) have examined approximate solutions obtained by restricting attention to the solution of the Schrödinger equation in one spatial dimension. If the control field operator U is assumed to be a multiplicative one (i.e. the control is an external applied potential), the dynamics being controlled reduces to the form

$$i\hbar\frac{\partial\psi}{\partial t} + \frac{\hbar^2}{2m}\frac{\partial^2\psi}{\partial\chi^2} - [V_0(\chi) + u(\chi,t)]\psi(\chi,t) = 0. \quad (7.2.21)$$

The function $V_0(\chi)$ is further chosen to be the Morse potential

$$V_0(\chi) = D[1 - e^{-v(\chi-\chi_0)}]^2 \quad (7.2.22)$$

which represents the potential energy function of a diatomic molecule. For this system of reduced complexity, numerical evaluations of the optimal control have been conducted in various cases of interest, such as when the applied field is spatially constrained (i.e. when $u(\chi,t)$ can be represented by the product $B(\chi)E(t)$ where $B(\chi)$ is a specified dipole function and $E(t)$ is a nonstationary electric field) and when the target state $\psi(\chi)$ is a Gaussian wave packet which is to be placed at various specified locations. By examining the characteristics of the solutions to the optimal control problem in these cases, it is demonstrated that the attained final state has a closer correspondence with the target state in the case when the fields are unconstrained than in the case of spatially constrained inputs, while requiring less energy in the function $u(\chi,t)$. As a further special case, a numerical solution to a formulation of the dissociation problem (breaking of specified molecular bonds) has also been obtained and optimal pulsing strategies that ensure dissociation in a specified time are developed. In an extension of these studies, the design of a robust controller that is insensitive to uncertainties in the molecular Hamiltonian and to errors in the initial state of the system is reported recently by Dahleh et al (1990).

Notwithstanding the successful demonstrations of the optimal field designs for quantum dynamical systems as discussed above, it should be noted that the solution of the Schrödinger equation and the numerical optimization required can become very computationally intensive when applied to practical problems. Active research is presently under way in the development of more efficient schemes (employing for instance parallel algorithms) to render these computations feasible. Much more work is needed in this area before polyatomic molecular systems can be routinely handled using these approaches. Nevertheless, the studies that have been conducted to date demonstrate the efficacy of employing tools from modern control theory for tailoring efficient strategies for control over molecular structure through organized approaches to achieve material modification at molecular scale.

7.3 Cosmic History of Time and C_{60}: New Moon Technology

This part of the chapter is written for three reasons: (1) the motivations for "serendipitously discovering the C_{60} molecule" were astrophysical for both Kroto and Huffman/Krätschmer; (2) Koruga became "awakened" to the importance of the C_{60} molecule for a "New World based on fivefold symmetry" upon reading Kroto's paper, "Space, Stars, C_{60} and Soot" (*Science*, 242, 1988) despite the fact that he had already read the Kroto/Smalley research team paper about C_{60} in 1985. This fact was a source for the third question--(3) If cosmic C_{60} is important for *space* and *stars*, what about *time*? Or, does a cosmic *spatio-temporal* icosahedron exist?

Kroto (Kroto et al, 1991) wrote "It is interesting to note that the motives for the experiments which serendipitously revealed the spontaneous creation and remarkable stability of C_{60} were astrophysical. Behind this goal lay a quest for an understanding of the curiously pivotal role that carbon plays in the origin of stars, planets, and biospheres. Behind the recent breakthrough of Krätschmer et al. in producing macroscopic amounts of fullerene-60 lay similar astrophysical ideas."

According to the Big Bang theory, stars began to form 15.9 x 10^9 years ago, the Earth 4.6 x 10^9 years ago, biospheres 2-3 x 10^9 years ago, etc. Considering the history of the cosmos and humanity, the following question arises: Does a *temporal icosahedron* exist as a law of cosmic history?

There is one indication for the existence of such a temporal icosahedron. The evolution of life, which started 3 billion years ago, may be viewed as a logarithmic spiral [one of the main properties of fivefold symmetry (Figure 7.3-1)]. Human technological innovation (from stone age to computer era) also follows a logarithmic spiral. Spirals, however, have *left* and *right* complementary orientations. What are the main events of cosmic complementary spirals?

For this, it is necessary to construct *the star pentagon*, which represents the Golden Mean Law. There exists *temporal synergy* (as synchronicity) between each part and whole of the star pentagon. A *temporal icosahedron* based on *left* and *right* spirals leads to the *heart model*. There are *five* "cosmic hearts," and each of them has six main events which are overlapping, giving three main points: *start-end*, *cross-section point*, and *end-start point* (Figure 7.3-2). According to this model, five further "temporal hearts" are possible: **[1] cosmic time is zero** (singularity--Universe as non-manifest mass)--**Universe becomes matter-dominated** (point "zero" in heart model). Double main event (point 0! in heart model): on the *left* spiral is particle creation and on the *right* spiral is radiation thermolysis. ***End-start*** point (point 1 in heart model) is the event of annihilation of electron-positron pairs. "Spiral 1" occurred for just 1 second, while "Spiral 2" occurred over 10,000 years. **[2] Dominant mass in Universe--planets' formation**, with double main event: birth of quasars (Spiral 1) and collapse of protosolar nebula (Spiral 2). *End-start* point (notation 1) is the formation of the population of stars. The "red shift" had a value of **10^{10}** [*red shift* is equal to $(1+v/c)/(1-v^2/c^2)^{1/2}-1$, where v is the velocity of the radiation source and c is the velocity of light--3 x 10^{10} cm/s], when cosmic time (from Big Bang) was *one* second. This event has the property of a center of inversion in symmetry, because (1) it happened about 10^{10} years ago, and (2) it was a preamble to the starting point for the cosmic origin of life in the form that we now know. [3] **Planets form -- origin of life** (molecular forms of life) with double event point: the formation of the Earth's atmosphere (Spiral 1) and the development of an oxygen-rich atmosphere (Spiral 2). The *end-start* point is the formation of water (seas and oceans where the first living organisms appeared). [4] **Origin of life--origin of human consciousness** with a double event: biological (chromosomes-DNA and centrioles-MT dynamical system association--Spiral 1) and human mind formation (self as source of unconsciousness based on mind-brain activity--Spiral 2). The *end-start* point is homo sapiens as a natural phenomenon (point 1 in heart model). [5] **Origin of human consciousness--origin of man-made machine consciousness** (*MMMC*) with a double event: psychological-spiritual one (before and after Christ--Spiral 1) and some event as a result of a crucial invention which will be based upon the C_{60} molecule (Spiral 2). The ***end-start*** point is a "twinkle" pentagon of C_{60} (in previous ***end-start*** point, small pentagons also exist) with main events: 1972 (Osawa C_{60} model); 1973 (Huffman: C_{60}-small particle production); 1978 (Koruga's carbon dream--unconscious penetration to consciousness); 1991 (mind synergy for C_{60} technology--preamble to C_{116} as C_{60} dimer); and 1996 (some important event, yet unknown). The center of the pentagon is 1985 (Kroto/Smalley team--the first Fullerene science pentagon) as initial pentagon of C_{60}. This consideration is presented as the "heart model and C_{60}-light machine" in Figure 7.3-3. In this consideration, Chinese (Kanji) ideograms were used to express the "*law of change*" through integration of human consciousness and unconsciousness. This new state of human mind [unification of the state of reality-consciousness ("awake") and the state of shadow-unconsciousness ("asleep") through dream ("lighting" of *Self* during sleep)] leads to a new technology based on the C_{60} molecule. The symbol which represents integration of human consciousness and unconsciousness is *the Moon* (as Chinese ideogram) and we will call the new technology of C_{60} based on this integration "*C_{60} New Moon Technology* (C_{60}-NMT)." A short explanation of Chinese/Japanese etymology of ideograms relevant to C_{60}-NMT is presented in Figure 7.3-4.

If the C_{60} molecule and fivefold symmetry have importance in cosmic evolution, then the *Holopent* (cosmic synergy of information as space-time pattern based on fivefold symmetry) and the *Cosmic Egg* become relevant. This was a subject of discussion between Koruga and Mojević (Koruga's friend and a well-known artist). Mojević created a painting with two eggs, a bigger one (space) and a smaller one (time) as one (information). The main properties of

F:7.3-1   **SPIRAL, SPACE-TIME EVOLUTION AND LIFE**

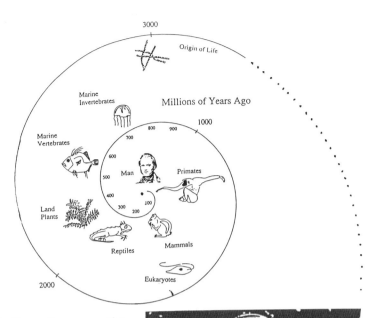

The Big Bang theory predicts that the Universe arose from a point. Life began eight billion years after "point explosion," or three million years ago. The evolution of life may be represented by spiral law. The evolution of human technology, which started about 10,000 years ago (starting with fire and wheel use) also has the property of a spiral.

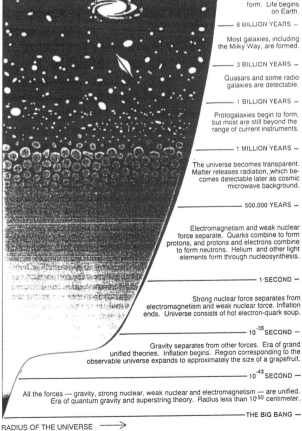

The sun and planets form. Life begins on Earth.

— 8 BILLION YEARS —

Most galaxies, including the Milky Way, are formed.

— 3 BILLION YEARS —

Quasars and some radio galaxies are detectable.

— 1 BILLION YEARS —

Protogalaxies begin to form, but most are still beyond the range of current instruments.

— 1 MILLION YEARS —

The universe becomes transparent. Matter releases radiation, which becomes detectable later as cosmic microwave background.

— 500,000 YEARS —

Electromagnetism and weak nuclear force separate. Quarks combine to form protons, and protons and electrons combine to form neutrons. Helium and other light elements form through nucleosynthesis.

— 1 SECOND —

Strong nuclear force separates from electromagnetism and weak nuclear force. Inflation ends. Universe consists of hot electron-quark soup.

— 10^{-35} SECOND —

Gravity separates from other forces. Era of grand unified theories. Inflation begins. Region corresponding to the observable universe expands to approximately the size of a grapefruit.

— 10^{-43} SECOND —

All the forces — gravity, strong nuclear, weak nuclear and electromagnetism — are unified. Era of quantum gravity and superstring theory. Radius less than 10^{-50} centimeter.

— THE BIG BANG —

RADIUS OF THE UNIVERSE $\longrightarrow$

space are pulsing and the relationship between part-whole and symmetry, while the main properties of **time** are its units ($1s=10^{10}$ of red shift) and sequential-parallel relationship (in the real world, present and past exist sequentially or event-by-event, while in the world of shadow, events are parallel, with no distinction between *past* and *future*; the present does not exist). The space-time pattern is information with the following major properties: replication, ordinal-cardinal (in the sense of mathematics, where "cardinal" refers to transfinite numbers) and Golden Mean. The result of this integration of the real world (Rw) (*past-present*) and the world of shadows (Sw) (*past-future*) will be *existence* as the **calm of to be**. For human beings, this will be possible if C_{60}-NMT solves the problems of human care and cosmic space-time.

Engineering without a biomedical perspective is blind, while biomedicine without engineering is helpless. The human **body** was the subject of research in the field of engineering in the 1950s. *Biomedical engineering* became a discipline at universities and engineering categories have become physiological terms. In the 1960s, the category of **mind** (intelligence) also attracted the attention of researchers in computer engineering, and a new engineering discipline--*artificial intelligence*--was born. During the same time period (1960s), the human brain was the subject of engineering research, but the concept of artificial intelligence (algorithmic, non-neural) won over neural networks (parallel, non-algorithmic) in the short range (Minsky-Rosenblatt duel). Like a phoenix, the concept of neural networks came onto the stage of engineering when Robert Hecht-Nielsen showed the first real-world applications of a computer designed by neural network principles in 1984. Until now, body, brain and mind have been engineering categories.

What is next? As there continue to be developments in computer science and engineering, we can expect *emotions* as a new engineering category. A *3H technology* has been proposed based on information physics as a link between Nature, Brain, Mind and Computers (Koruga, 1992). The 3H technology is meaningful in the realization of information machines (computers) in their three phases of development. The first phase encompasses *Hand* technology and is represented by computers with which we communicate by keyboard--by hand (first "H")--today's computers. The second "H" phase--Head--includes those which will be autonomic intelligent information machines, capable of movement in space and time and of adapting to their environment, realized on the principles of brain functions. The first generation of such machines has already been built utilizing artificial neural networks, with the

F:7.3-2

DEUS EX MACHINA
Temporal icosahedron - Heart model - C_{60} molecule

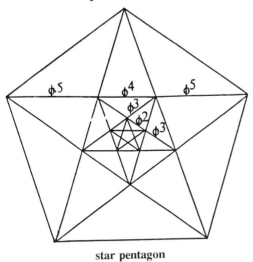

star pentagon

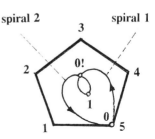

spiral 2   spiral 1

Temporal icosahedron
as Heart model. Each "heart"
has three main points:
0 (start-end); 0! (double
main event); and 1 (end-start)

The star pentagon represents the ratio of the diameter to the side of a pentagon which through the diagonals gives the value of the golden mean. The biological structure (microtubules) through cylindrical representation of a pattern has $\phi^{-2}[\phi(\bar{2})_5]$ and content information of shadow world (Sw) as $N=(\bar{2})_5$. The history of time (Big Bang) also can be represented as a temporal icosahedron (fivefold symmetry). In time, the history value 10^{10} of red shift represents the cosmic time of one second.

Major events in the universe's history which are important for life and technology are:

1. Universe as non-manifest mass, cosmic time is zero, epoch is singularity ($N=0$), red shift is infinite, it was 2.0×10^9 years ago. The first spiral happened in 1 second, when in that period annihilation of electron-positron pairs was happening (all was light). A symmetrical spiral to the previous one happened in the next 10^4 years, and it finished when the universe became matter-dominated.

2. A new spiral represents the period from dominant mass in the universe to the first stars' formation (4.1×10^9 years from the Big Bang--BB). A symmetrical spiral to this started from stars' formation to planets' formation (15.4×10^9 years from BB).

3. A third pair of spirals started from planet formation through water formation on Earth to the origin of life (19×10^9 years from BB).

4. The spiral of evolution started from the origin of life (macroscopic life forms) and finished with *homo sapiens* (20×10^9 years from BB or 1×10^5 years ago). A symmetrical spiral to biological evolution is *homo sapiens'* evolution of unconsciousness, which started 100,000 years ago and finished about 10,000 years ago.

5. A fifth pair of temporal icosahedron spirals started about 10,000 years ago, when *homo sapiens* became self-conscious and started to create (stones, fire, the wheel, etc.). The first spiral finished in the 1970s, and a symmetrical one was started, when fivefold symmetry and the C_{60} molecule became the subject of scientific consideration. A new spiral of cosmic time, which should close a temporal icosahedron as fivefold symmetry, is science, technology and engineering based on the C_{60} molecule.

F:7.3-3

HEART MODEL AND C$_{60}$-LIGHT MACHINE

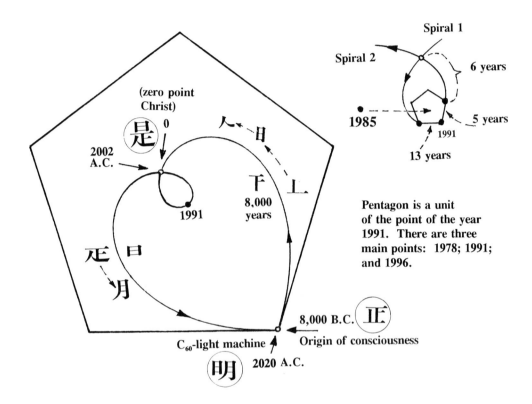

We usually say that human civilization started 10,000 years ago (ancient Egypt, India, China, etc.), when something changed ("big switch") in the human mind, from an unconscious state to a self-conscious one, from a passive to an active interaction with nature. *Stone* was the first tool in human creative interaction with nature. *Fire* was one of the first secrets which man stumbled upon accidently from nature. The *wheel* was his first serious invention, and so on . . . . The human being became self-conscious (正 → ¯ + ⼂ + 日 → 人) and his mind through interaction with nature "growth" and "growth." There were a lot of important events in human history but one, for his mind overcame all of them when he/she split history into two parts: *all era* and *new era* (before and after Christ). This event split the first spiral into two parts and it is noted as 0 ("zero"). The second part of the first spiral is from "zero" to the year 1991, when the C$_{60}$ molecule became the subject of the unified human mind (unconscious and self-conscious states). The new spiral, symmetrical to the previous one, started in 1991 and, according to the "heart model," will have an important event in 2002 (a new machine based on C$_{60}$) as a joint point with the "zero" event of the previous spiral. The preamble to 1991 was the year 1978 ("carbon dream" - 13 years), and the postamble is 1996 as some important event. According to the "heart model" (Cosmic Egg symmetry model), this spiral will finish about the year 2020, when technology and engineering based on the C$_{60}$ molecule will become a dominant cosmic "light machine," which will reversibly transform human information directly to machine and vice versa.

ultimate achievement of this generation being some kind of *awareness* as a preamble to the "artificial" consciousness of the next generation. The third phase, *Heart*, will consist of computers possessing elements of "artificial" consciousness and "artificial" emotions, on the basis of which they will be able to self-adapt relations amongst themselves and between them and their environment, including their relations with human beings. We believe that C_{60}-NMT will be one of the main technologies that will be based on Head-Heart principles and will be capable of solving the *human care* and the human space-time cosmic problems.

F:7.3-4

C$_{60}$ NEW MOON
Synergy challenge to Fullerene technology

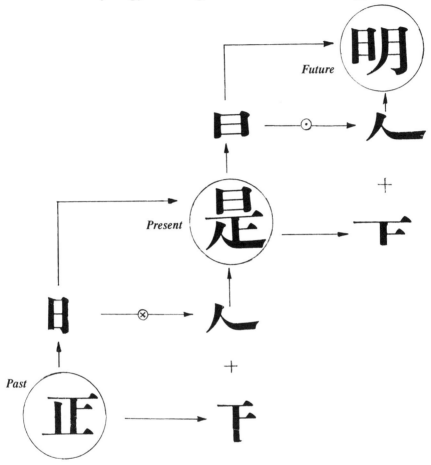

Chinese ideogram (Kanji) 正 means *correct*, but it also means *perpendicular*. These two words are synonyms in the Chinese and Japanese languages. This ideogram has five strokes, but three basic elements (one and two structures which are perpendicular). One of these structures means divination (卜). Another part (⊥) from sun (日) changes to 人 (man). A new form of the same character is 是, which again means correct (Japanese) and represents the verb *to be* (Chinese). Univication 人 (human) and 卜 (divine) as new one through 日 (sun) gives 月 (moon). This is a New Moon, which together with the sun gives brightness (明), which is the basic element for further characters: 明日 (Japanese) and 明天 (Chinese). These three ideograms 正 , 是 and 明 have the same meaning in temporal icosahedron. New Moon is the concept of unity of our unconsciousness and consciousness in new technology realization. The C$_{60}$ molecule was the subject of astrophysical research but the main contribution of this molecule will be in new machine concept and a new mind (unification of our consciousness and unconsciousness).

COSMIC EGG
Holopent and C_{60} molecule

1. **INFORMATION**
    Replication
    ordinal-cardinal
    Golden mean

2. **SPACE**
    Pulsing
    part-whole
    symmetry

3. **TIME**
    $1[s] = 10^{10}$ of red shift
    sequential-parallel
    temporal icosahedron (Heart model)

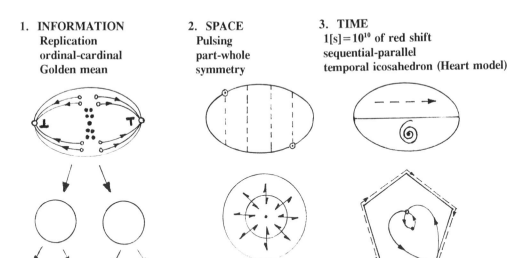

The Holopent is a synergy of information-space-time structures based on fivefold symmetry (from the Greek *holos*, meaning entire, and *pente*, meaning five). The C_{60} molecule possesses fivefold symmetry (icosahedron) and could be one of the basic elements of this "Cosmic Egg" in its third stage (Nature-Living Being-Machine). *"It is interesting to note that the motives for the experiments which serendipitously revealed to spontaneous creation and remarkable stability of C_{60} were astrophysical."*--Kroto.

Artist's vision of the Cosmic Egg. Dragan Mojević (an internationally known artist) [left] and Djuro Koruga [right] in Mojević's atelier (1989), talking about the Cosmic Egg, the Holopent, and the relationship of art and new technology based upon fivefold symmetry.

PART III

FULLERENE C$_{60}$: PRODUCTION TECHNOLOGY AND APPLICATIONS

The scenario events of Universe are the regenerative interactions of all others and me. Universe is the starting point for any study of synergetic phenomena. Universe is the minimum of inter-transformings necessary for total self-regeneration.

– Buckminster Fuller

CHAPTER 8

FULLERENE PRODUCTION

8.1 General Approach

When Smalley's team (Kroto et al, 1985) observed the existence of C_{60} in 1985, the race was on to synthesize the truncated icosahedron structure to provide material for characterization and applications development. Most early synthesis investigators attempted the more traditional organic synthesis approaches which were unsuccessful. Not until the Huffman and Krätschmer (Krätschmer et al, 1990a, 1990b; Huffman, 1990a, 1990b) discovery was the existence of Fullerenes confirmed and for the first time C_{60} was proven to be that showed a geodesic ball structure consisting of 20 hexagonal and 12 pentagonal faces. The Huffman/Krätschmer (H/K) synthesis consisted of vaporizing carbon from a graphite rod in a residual atmosphere of helium at a pressure in the 100-200 torr range. The graphite rod vaporization can be accomplished via resistance heating of the rod or producing an arc between two rods using either **ac** or **dc** power. Carbon soot, produced by condensation of the vapors collected and dissolved in an organic solvent. Preferred solvents are benzene ring compounds such as benzene, toluene, tetramethylbenzene, methylnapthalene, etc. The Fullerenes are soluble in many organic solvents are separated from the non-Fullerene soot using filtration. The organic solvent is volatilized leaving Fullerene crystals.

Although most anyone who has access to an airtight container that can maintain a helium pressure in the 10-200 torr range and that has electrical feedthrough to deliver power to vaporize carbon, can produce minute quantities of Fullerenes. However, if multi-gram quantities are desired at a reasonable price, more sophisticated equipment and processing are required to economically produce kilogram quantities. Although the mechanism of Fullerene formation has been suggested by Smalley (Curl and Smalley, 1991; Smalley, 1992) and others (Diederich and Whetten, 1992), experimental processing has not demonstrated confirmation of the hypothesized mechanisms to the extent that process controls result in predictable reaction kinetics. Since MER is the only licensee of the Huffman/Krätschmer intellectual property and have commercially supplied Fullerenes since December 1990, considerable process variations have been demonstrated that affect the synthesis of Fullerenes.

Huffman and Krätschmer utilized nominally 3 to 6 mm diameter graphite rods that were vaporized in a glass ball jar by resistance heating or arcing. The collected soot contained a few percent benzene soluble Fullerenes. Typically only a few to less than ten milligrams of Fullerenes were produced.

F:8-1   **MER Block Diagram of C$_{60}$ Production**

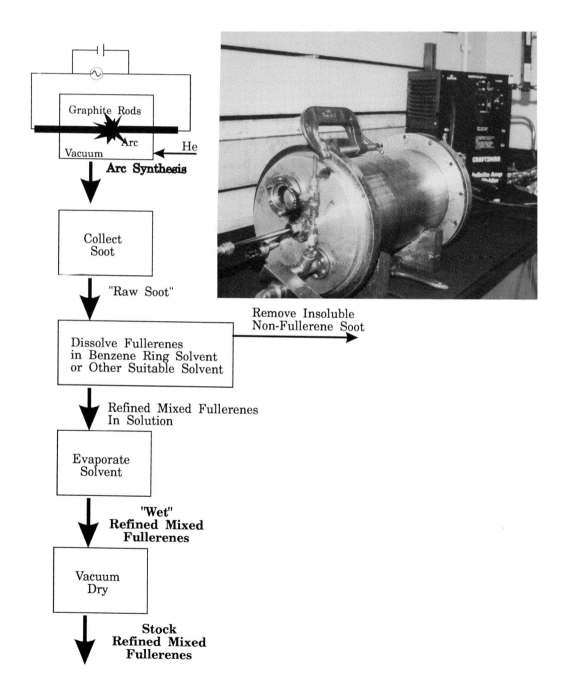

PRODUCTION OF C$_{60}$

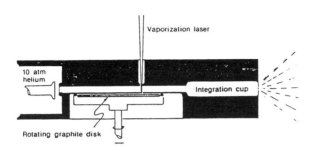

Schematic diagram of the pulsed supersonic nozzle used to generate carbon cluster beams. The vaporization laser beam (30-40 mJ at 532 nm in a 5-ns pulse) is focused through the nozzle, striking a graphite disk which is rotated slowly to produce a smooth vaporization surface. Helium carrier gas provides the thermalizing collisions necessary to cool, react and cluster the species in the vaporized graphite plasma, and the wind necessary to carry the cluster products through the remainder of the nozzle.

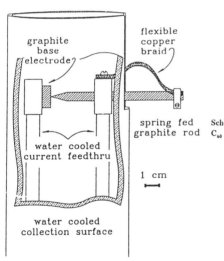

Schematic diagram of graphite rod contact-arc C$_{60}$ generator.

FULLERENE FACTORY makes macroscopic samples in a carbon arc. The arc--a refinement of an apparatus developed by Wolfgang Krätschmer and Donald Huffman--frees carbon atoms that coalesce into sheets. Inert helium holds the sheets near the arc long enough for them to close in on themselves, forming fullerenes.

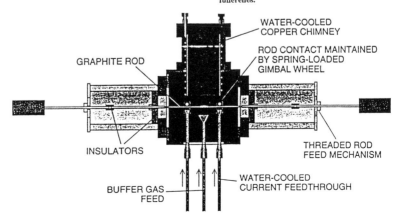

An example of a bench scale batch system of this nature is shown in Figure 8-1. A schematic of the system Huffman originally utilized in the discovery (Krätschmer et al, 1990; Huffman, 1990) is given in Huffman (1991). Clearly a more mechanized system was necessary to produce multi-gram quantities. Smalley (Curl and Smalley, 1991) illustrated a system based on the principle of that shown in Figure 8-2, with a mechanized threaded feedthrough to maintain an arc as one or both graphite rods are consumed. While such a system increased the production of soot by continuously burning one (**dc** power) or two (**ac** power) rods up to approximately 30 cm long, it does not produce large quantities of soot. Depending on the quality of the graphite rod, a 30 cm rod weights approximately 15-18 grams and with a high conversion of rod to soot only 7 to 25 or 30 grams of soot is produced per run, which consumes several hours of the operations that consist of burning the rod, reloading, pumping out the air, and preparing for a repeat sequence results in only limited quantities of Fullerenes.

D. H. Parker et al. (Parker et al, 1992) surveyed the methods used to synthesize of Fullerenes. The methods utilized to evaporate carbon were resistance heating (Taylor et al, 1990; Aije et al, 1990; Cox et al, 1991; Hare et al, 1991), **ac** or **dc** arc discharge (Haufler et al, 1990; Haufler et al, 1991; Whetten et al, 1991; Bae et al, 1991; Koch et al, 1991; Pang et al, 1991; Parker et al, 1991; Shinohara et al, 1991), laser ablation (Kroto et al, 1985; Curl and Smalley, 1991; Smalley, 1992; Haufler et al, 1990), and high frequency induction heating (Peters and Jansen, 1992). Withers (Withers and Loutfy, 1992) has used additional carbon evaporation methods to vaporize carbon which has included: electron beam including sputtering, microwave, plasma arc, inductive plasma, and a combination of these vaporization processes including with a burner. Howard (Howard et al, 1991) at MIT was the first to demonstrate that combustion of benzene under similar pressure conditions to those of the Huffman/Krätschmer method could yield Fullerenes. Withers (Withers and Loutfy, 1992) has used other carbon precursors to burners that yield Fullerenes. The precursors include carbon particles or soots and virtually any hydrocarbon with a hydrogen to carbon ratio less than one. A two stage process to dehydrogenate the hydrocarbon precursor can be used and burners operated at atmospheric pressure can also produce Fullerenes. As stated above, combination processes can be used with burners such as microwave and inductive plasmas to produce Fullerenes. These processes, which are continuous, have the potential for large scale production of Fullerenes.

Most synthesis of Fullerenes through 1992 has utilized the Huffman/Krätschmer processing or arcing graphite rods. While soot containing Fullerenes is relatively easy to produce, there are a number of processing variables that dramatically affect the quantity of soot per unit time, the quantity of Fullerenes within the soot and the molecular weight distribution of Fullerenes. Experience has shown that the chamber in which the arcing takes place must have high vacuum capabilities. Although the arcing typically takes place at about 100-200 torr, it is important that the oxygen concentration be kept very low during vaporization thus dictating a well sealed chamber that permits pumping down to at least 10^{-3} torr followed by backfilling with helium. Residual oxygen in the chamber or leaks in the feedthroughs can depress the Fullerene concentration in the soot to only a few percent to virtually zero, from possible yields of 10-20%. This sensitivity to oxygen makes a batch process very difficult to produce high yields, since the system must be opened after each rod is burned to put in new rods and harvest the soot. In addition to the absence of oxygen, the residual gas has a dramatic effect on soot, Fullerene yield and molecular distribution weight of Fullerenes produced. If inert gases other than helium are utilized, little to no soot is vaporized from the carbon arc. Although argon will produce Fullerenes, the yield is substantially less than with helium. In nitrogen, no Fullerenes were produced. In mixtures of helium and argon, the relative amounts of C_{60} and C_{70} produced could be influenced. Hydrocarbons with a low hydrogen to carbon ratio as the residual gas can improve the yield of Fullerenes.

The type of power supply can have an effect on the quantity of soot produced. Some have reported that **dc** arcing is more efficient than **ac** (Pang et al, 1991; Shinohara et al, 1991), while others (Walton and Kroto, 1992) report no difference. The MER research team has not found any difference on the average for a large number of runs, between **ac** and **dc** for the percentage of Fullerenes in the soot. However, **ac** produces a larger quantity of soot than **dc**. From an **ac** arc 80 to over 90% of the rod weight can be transformed to soot, while at the same power conditions only about 60-65% of the rod is transformed to soot using dc arcing that will yield equivalent soot to ac arcing. The percentage of Fullerenes in the soot is about the same and in MER's case about 8-30% is the range under production conditions with typical average values of 12-15%. Values as high as 40% and up to 50% or more toluene extractable Fullerenes in soot for individual runs have been obtained which provides a target of what can ultimately be achieved. Higher molecular weight solvents than toluene have been reported up to 94% Fullerenes extracted from soot (Parker et al, 1992).

The size of graphite rod arced to produce soot has an effect on soot yield and more importantly on Fullerene yield. Huffman utilized ⅛ and ¼ inch diameter graphite rods. Withers utilized ¼ inch graphite rods, but when larger rods such as ⅜, ½ or ¾ inch diameter rods were used both the soot and Fullerene yield decreased. At constant current density for larger rods, the yield was reduced which suggests arc temperature profile plays an important role in the yield of Fullerenes. Koch (Koch et al, 1991) found that lower currents gave better yields. Pradeep et al (Pradeep and Rao, 1991) reports primarily C_{60} is produced when using less than 150 amps on a 5mm graphite rod. The MER research team has investigated a wide current distribution from about 70 amps to about 250 amps on a ¼ yield (6.35 mm) rod. In general, the lower the current the higher the yield both of soot and toluene-soluble Fullerenes in the soot. However, production rate of soot is impractically low at under 100 amps. There is also a relationship between residual helium pressure and current on the soot production as well as Fullerene yield. Thus there is a compromise of rate of soot production, soot yield and Fullerene yield no matter whether ac or dc current is used.

The quality of graphite has some effect on both the soot and Fullerene yield. Pang et al. (Pang et al, 1991) produced Fullerenes from demineralized coal. Withers has investigated all available graphite rods and found that the higher the quality of graphite, the better the yield, although there is not a great deal of difference in yield. The MER research team utilized reclaimed carbon from automobile tires (Withers and Loutfy, 1992) and a variety of low quality cokes and pitches that were adequate for precursor carbon to synthesize Fullerenes. The spent soot, after separating the Fullerenes, is an excellent precursor carbon that provides higher yields of Fullerenes in the soot produced (Withers and Loutfy, 1992). It appears practically any carbon precursor will yield Fullerenes provided it does not contain excessive impurities, particularly oxygen or hydrogen. However, oxygen as a metal oxide does not seem to adversely affect Fullerene formation. In fact, Smalley (Smalley et al, 1991), Withers, and many others have doped graphite rods with metal oxides which are reduced in the arc or laser ablation technique to encapsulate the metal inside the Fullerene and presumably produce CO. Such endohederal complexes of metals inside the Fullerene cage have a variety of potential applications, and the product can be synthesized by doping metal oxides in the graphite rod as well as introducing oxides directly to the arc.

The arc itself has an effect on the yield of soot and Fullerenes. The arc distance can be varied from about ½ to almost 10 mm for a ¼ inch diameter graphite rod. In general, Withers has found that a gap of about 2mm is the most desirable for producing high yield soot and Fullerenes. Most batch systems are hand fed or screw fed which is difficult to control the gap distance. Withers developed an automatic computer controlled system in mid-1991 that has the capability to precisely control gap distance, current and pressure which optimizes soot and Fullerene yield. Rod feed is automatic and several dozen rods can be burned without opening the system which maximized soot production and Fullerene yield per unit time.

Since in the case of a dc arc, only the anode is vaporized, it is possible to utilize particulate carbon and feed continuously (Withers and Loutfy, 1992). Any suitable carbon particulate can be fed continuously to the arc as the anode which will be vaporized to produce Fullerene containing soot. The cathode rod is not consumed which permits the operation of a continuous process utilizing the most economical form of solid carbon - particulate, fed continuously as an anode (Withers and Loutfy, 1992).

Regardless of how the soot containing Fullerenes are produced, it is necessary to separate or extract Fullerenes from the raw soot and to finally fractionate into specific molecular weights. The separation or extraction of Fullerenes from raw soot is readily accomplished utilizing the solubility of the Fullerenes. As previously mentioned, Fullerenes have limited solubility in a very wide variety of solvents, but toluene is the most often used solvent for extraction. Parker (Parker et al, 1992) found that Soxhlet extraction worked much better than simple reflux, giving a 26% yield of soluble material compared to 14% by reflux. Withers found Soxhlet extraction was slightly worse than room temperature toluene extraction provide the solubility of the toluene was not exceeded. Also, room temperature extraction could be accomplished in about ½ hour compared to a 24 hour Soxhlet extraction. The toluene extracts primarily Fullerene molecular weights of less than about C_{100} as reported by Parker (Parker et al, 1992) and similar results found by Withers. Higher boiling point or molecular weight solvents will extract higher molecular weight Fullerenes from the soot. Solvents such as tetramethylbenzenes or methylnapthalene will extract molecular weights up to about C_{450}. The quantity of Fullerenes above C_{100} can be up to about the same as below C_{100} depending on the synthesis processing conditions. Parker (Parker et al, 1992) reported about 26% of toluene soluble Fullerenes and 44% in 1,2,3,5-tetramethylbenzene. N-methyl-2-pyrrolidinone (NMP) extracted 94% of the raw soot which showed molecular weights as low as C_{30} . NMP provided higher solubilities from the soot probably because of increased solvent penetration breaking bonds that hold the extractable Fullerenes within the soot since NMP is known to penetrate coal structures, and separate olefines and arcomatics for refining oils. High temperature high pressure bomb extraction using toluene is reported to extract a greater quantity of Fullerenes from the soot and higher molecular weight Fullerenes (Lamb et al, 1992). Parker (Parker et al, 1992) was unable to repeat these results and Withers found only a small difference between high temperature high pressure bomb extraction and room temperature extraction using toluene. This indicates the sensitivity and lack of understanding for both the synthesis and extraction of Fullerene since one group can observe substantially different results from another. This is probably due to all experimental conditions not being equivalent.

8.2 Separation

The application of Fullerenes requires separation into specific molecular weights as well as molecular weight fractions. Chromatography methods have proven most successful and were used originally by Huffman and Krätschmer (Krätschmer et al, 1990a, 1990b; Huffman, 1990a, 1990b). A variety of stationary phases have demonstrated success including silica gels, silica gels with absorbed phases, carbon blacks and aluminas. The neutral aluminas have proven the most successful as the column packing. Either extracted Fullerenes or soot can be used and eluted with practically any solvent that dissolves the Fullerenes. It is desirable to utilize a solvent that has high solubility of the Fullerenes and as much solubility difference between the molecular weights. By utilizing differential solubility of the solvents to molecular weight of the Fullerenes, the task of fractionation can be made easier. For example, hexane will dissolve primarily only C_{60} and C_{70} from the soot and toluene $< C_{100}$. To differentiate solubility between molecular weights, mixed solvents are helpful and can be used in parallel columns to effectively separate C_{60}, C_{70}, C_{84}, etc. Withers utilizes this approach to commercially produce fractions of C_{60} 99.9 + % pure, C_{70} 99 + % pure, C_{84} 98 + % pure, etc. The fraction between C_{84} and C_{100} if a toluene extract was used, is the molecular weight fraction range which can be utilized for applications requiring higher molecular weights. If a fraction of $> C_{100}$ is desired, then after a toluene extraction of the soot, tetramethylbenzene or methylnapthalene can be used to obtain a fraction of about C_{100} - C_{240}. If molecular weights higher than C_{240} or less than C_{60} is desired, the NMP solvent can be used for soot extraction. Thus through using

F:8-3

Polyhedral Shapes in High Current Pyrolytic
Deposits Formed on the Cathode at 550 Torr

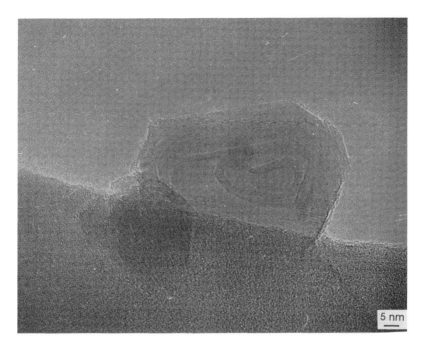

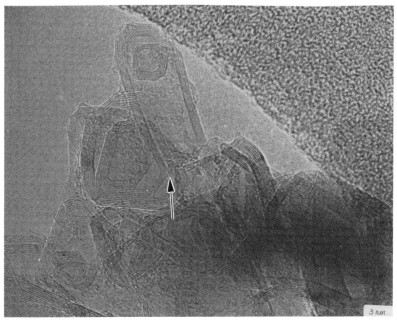

selected solvents for soot extraction and chromatography, it is possible to produce very pure Fullerenes of a specific molecular weight or molecular weight fractions.

The Fullerenes in contrast to other forms of carbon can be sublimed in a vacuum at relatively low temperature. The sublimation point is also related to the molecular weight of the Fullerene. It is thus possible to fractionate the Fullerenes with careful temperature and pressure control using sublimation. No one has yet reported utilizing a system with sufficient control to produce high purity fractionated Fullerenes. However, extensive thin film work is carried out with evaporated Fullerenes.

8.3 Nanotubes

As mentioned above, residual pressure, arc distance and power can have a dramatic effect on the Fullerene produced. Also mentioned above, with dc arcing a pyrolytic "slag" is produced on the cathode which reduces the amount of soot produced. Iijima (Iijima, 1991a; Iijima, 1991b) first revealed this solid pyrolytic slag on the cathode of a dc arc contained tubular shapes which are referred to as nanotubes or "buckytubes".

Subsequently, Ebbesen and Ajayan (Ebbeson and Ajayan, 1992) reported synthesis conditions for nanotubes are improved with increasing pressure. The best pressure for buckyball synthesis is 100-200 torr and for Buckytubes Ebbesen and Ajayan (Ebbeson and Ajayan, 1992) reported 500 torr to be the best pressure. Withers found pressure is important but tube formation is more sensitive to power. Using a ¼ inch (6.35 mm) diameter graphite anode rod, a current higher than about 100 amps produces more polyhedral shapes such as shown in Figure 8-3.

A low current and controlled gap in an automatic computer controlled system produces excellent tubes such as shown in Figures 8-4 and 8-5. A pyrolytic "slag" deposit shown in Figures 8-6 and 8-7 contains tubes in both the core and shell sections of the pyrolytic deposit shown in Figure 8-8. The tubes are 5 nm to 60 nm in diameter containing two to fifty graphite shell layers and are up to 10μ (10,000 nm) long. Work by the MER research team is in progress to produce longer and even continuous tube shapes and identify the electrical and mechanical properties of both the tubes and polyhedral shapes. Buckytubes and C_{116} (as a dimer of C_{60}) have only recently been discovered and offer an even greater potential than buckyballs, in pure form, in solution, and in a thin film.

F:8-4

Nanotubes (Buckytubes) Formed at Low Current and 550 Torr

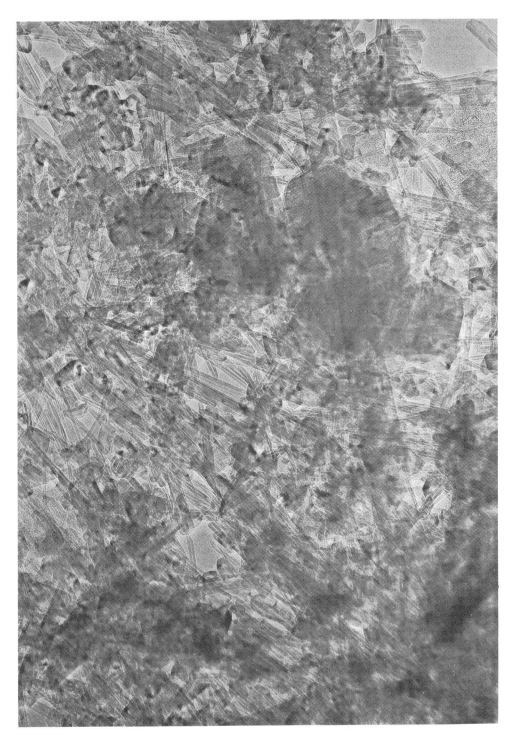

F:8-5

Nanotubes From an Automatic Computer Controlled System

F: 8-6 and 8-7   **Pyrolytic Slag from Cathode that Contains Buckytubes**

Core Section of Pyrolytic Carbon Slag Deposit

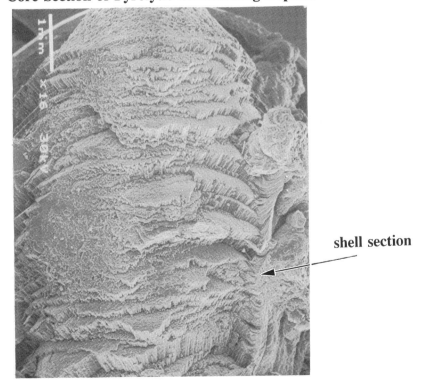

shell section

F:8-8

Cross Section of Pyrolytic Slag

Polyhedral Shapes in High Current Pyrolytic
Deposits Formed on the Cathode at 550 Torr

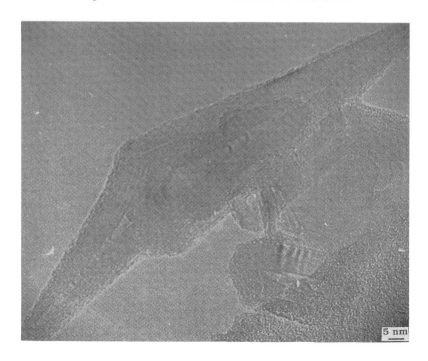

CHAPTER 9

NANOTECHNOLOGY OF C_{60}

9.1 What is Nanotechnology and Where is it Going?

One general definition may be given as follows: **Nanotechnology** is a new scientific and engineering paradigm which provides for the manipulation of atom(s) and molecule(s) for building new materials and/or molecular devices. The basic concept of nanotechnology was first proposed in 1974 by Toniguchi of Japan. Nanoscale fabrication principles were proposed later by Drexler (1981), and by Schneiker and Hameroff (1987). One of the ultimate goals of nanotechnology is to build nanoscale devices that allow electrons to pass through a circuit one-by-one at room temperature. One of the principles of nanoscale devices will be a combination of the Coulomb interaction between electrons and their passage by quantum tunnelling through an insulating barrier (Devoret et al, 1992).

An important component in nanotechnology research is scanning tunnelling engineering (STE). In numerous experiments, it has been shown that using an STM technique, the tip may be used to modify the surface of the sample in different ways, such as drawing fine lines to provide nanometer lithography. A surface can be modified by changing the STM voltage from its normal imaging level of 1 V to 5 V. This can, for example, produce a feature on the C_{60} molecule (e.g. the removal of atoms to make a smaller Fullerene) or effect internal doping. Another nanotechnology method is self-assembly of molecules. A combination of STM modification and self-assembly may provide the best solution. In accordance with our special interest in Fullerene C_{60}, we will present both self-assembly of "weak" dimers of C_{60} and scanning tunnelling engineering of "strong" dimers of C_{60}.

9.2 "Weak" Dimer of C_{60}

Weak dimers are two C_{60} molecules joined by a single carbon-carbon bond. This approach was proposed by Morton et al (1992) and is presented in Figure 9.1-1. For molecular computing based on non-linear oscillatory processes and self-assembly mechanisms (RC_{60} $C_{60}R$ ⇆ 2 RC_{60}), a "weak" dimer is very promising. For self-assembly of "weak" dimers in a 2-D lattice, platinum, palladium and nickel chemistry may be used. For a 3-D lattice like a biological microtubule, 4.4-bis pentanol could be used as the building block

F:9.1-1

"WEAK" DIMER OF C_{60}

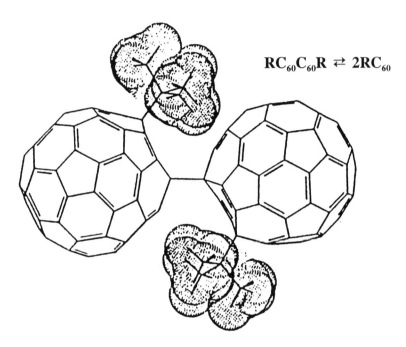

$$RC_{60}C_{60}R \rightleftarrows 2RC_{60}$$

Molecular model of the dimer [tert-butyl-C_{60}]$_2$

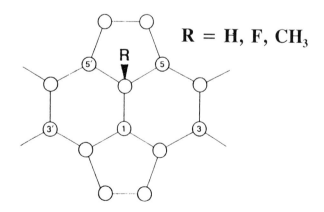

$$R = H, F, CH_3$$

Monoalkyl radical (R) attached C_{60} on double bonds

for a "weak" metallic complex C_{60} dimer ($wC_{60}dMC$). These macromolecules create cylindrical structures in the self-assembly process, and hexagonal packing on the cylindrical surface of the "$wC_{60}dMC$" is possible as artificial C_{60} microtubules ($aMT-C_{60}$).

9.3 "Strong" Dimer of C_{60}

A strong dimer (C_{116} as a dimer of C_{60}) was proposed by Koruga and published in the journal *Fullerene Science and Technology* (Koruga et al, 1993). The structure of the C_{116} molecule is presented in Figure 9.2-1. The starting material consists of two C_{60} molecules; then a C_{58} molecule is made as an open structure of C_{60}. Each C_{58} molecule has four free bonds for interaction with other C_{58} molecules. When two C_{58} molecules join together, they make a dimer C_{116}.

The highest occupied molecular orbital (HOMO) and lowest unoccupied molecular orbital (LUMO) of the C_{58} molecule have been calculated and are presented in Figure 9.2-2. The energy necessary for the dissociation process $C_{60} \rightarrow C_{58} + C_2$ is 5.2 eV. The energy value per atom of C_{58} is 1.63 times higher than in the C_{60} molecule, while the C_{116} molecule has an energy value per atom just higher, at 0.009 eV. The electronic structure and hybridized molecular orbitals of the C_{116} molecule are calculated and presented in Figure 9.2-3.

A similar procedure to that used for imaging Fullerene C_{60} with atomic resolution was also used for the nanotechnology of C_{60}. First, a 9x9 nm surface with C_{60} molecules in the "frozen" state was imaged. C_{60} was then imaged to look for irregularities (defects). STM images showed a regular structure, without defects. (Figure 9.2-4: bias voltages were 0.0201 V and 1.0 nA). Then two C_{60} molecules in the "frozen" state were exposed for less than 10^{-2} seconds to high voltage (5.2 eV). Subsequently, an STM image showed two C_{58} as open C_{60} molecules. The picture of C_{58} which is the result of computer calculation (molecular orbitals--Figure 9.2-1) and the images of the C_{58} molecules that were obtained with STM (Figure 9.2-5) are very similar or the same, strongly suggesting that the imaged structures are C_{58}.

It is well known that Fullerenes rotate and move around their substrate in normal situations, both in solution and in the crystal state. When the "frozen state" was "unlocked," two C_{58} molecules apparently rotated and moved, joining together spontaneously to make a C_{116} molecule as a "strong dimer." When the "frozen state" was again established, an STM image showed the

F:9.2-1

C_{116} DIMER OF C_{60}

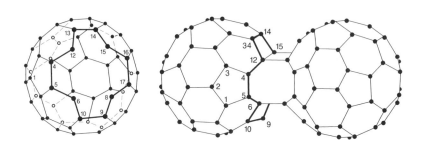

Front view of the shape of C_{58} as open structure of C_{60} molecule. When two open C_{60} molecules (C_{58}) join together, then make dimer C_{116}.

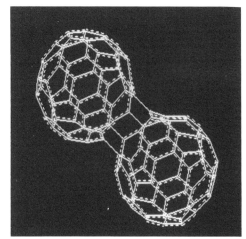

Molecular orbitals of C_{58} as open structure of C_{60}. Calculation is based on HyperChem program. Stick model of C_{116} molecule as dimer of C_{60}. Four links between two monomers are visible.

F:9.2-2

C_{58} AS OPEN STRUCTURE OF C_{60}
HOMO - LUMO

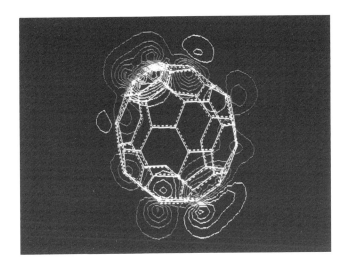

The highest occupied molecular orbital (HOMO) of C_{58} molecule as open structure of C_{60}. When we compare this picture with HOMO of C_{60}, the asymmetry of molecular orbitals and energy are recognizable.

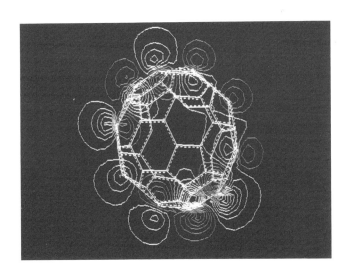

The lowest unoccupied molecular orbital (LUMO) of C_{58} molecule as open structure of C_{60}.

F:9.2-3

C$_{116}$ MOLECULE

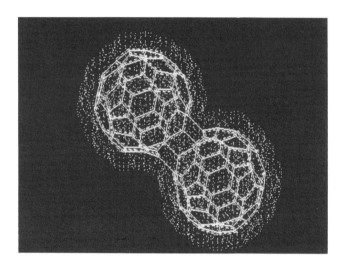

Electronic structure of C$_{116}$ molecule. Positions of carbon atoms and electronic cloud of molecule are visible.

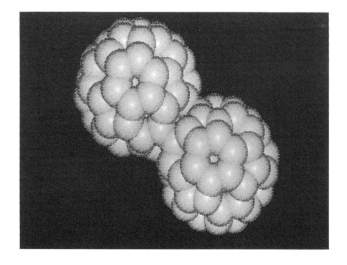

Hybridized molecular orbitals of C$_{116}$ molecule. Positions of pentagons and hexagons are visible. Hexagons possess an "energy hole," while pentagons do not.

structure of the Fullerene as that of a C_{116} dimer.

STM images of a C_{116} molecule (Figure 9.2-5) differ from a computer simulation of a C_{116} molecule. The STM image shows two "holes" in the surface of the molecule, while computer simulation of electron density shows a closed surface and a picture of hybridized molecular orbitals shows two small holes (Figure 9.2-3). It is a situation similar to that of pictures of the C_{60} molecule in solution and in the crystal state (Figure 3.2-4). The C_{60} molecule in the crystal state has a "hole," while in solution it has a closed surface. This indicates that the properties of solid C_{60} (crystal state) and the C_{116} molecule (as a dimer of C_{60}) may be similar. When the C_{116} dimer can be synthesized (chemically) in sufficient quantities, more of its chemical and physical properties will be characterized. One of its very important characteristics can be predicted: C_{116} rotates in one direction and has an oriented magnetic field. As we know, this is not the case with the C_{60} molecule (which rotates randomly and has a magnetic field which is not oriented). This property of the C_{116} molecule may open a new "page" in Fullerene technology history, with applications in superconductivity, quantum devices and molecular electronics.

F:9.2-4

NANOTECHNOLOGY OF C_{60}

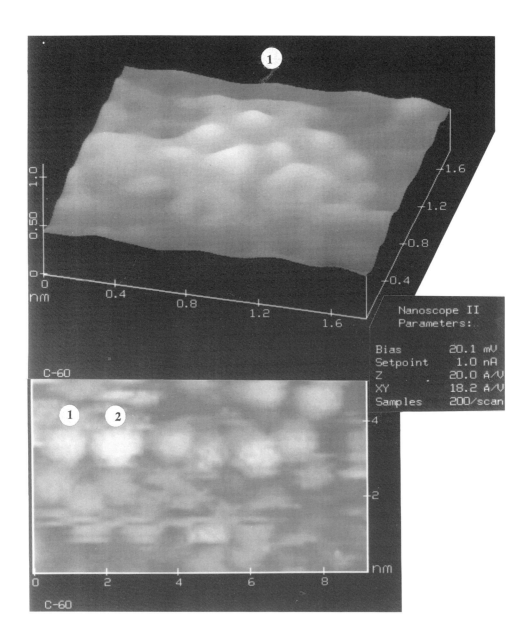

STM image of C_{60} molecule on gold substrate. Two C_{60} molecules (1 and 2) were the object of our research (lower picture). The internal structure of one of them (1), with pentagons (5 + one in middle), is visible (no defect).

F:9.2-5

NANOTECHNOLOGY
C_{60} - C_{58} - C_{116}

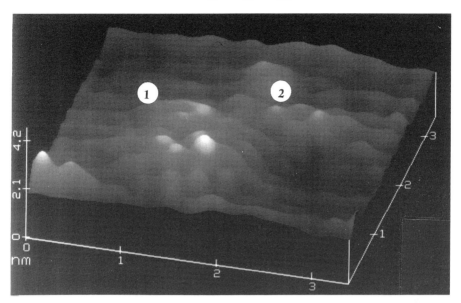

Two C_{58} molecules (1, 2) as open structure of C_{60}. Two carbon atoms are removed. C_{58} molecules are fixed (no-rotation, -sliding, -bouncing). Four places with higher energy state are visible (this also was predicted by computer simulation of C_{58}.

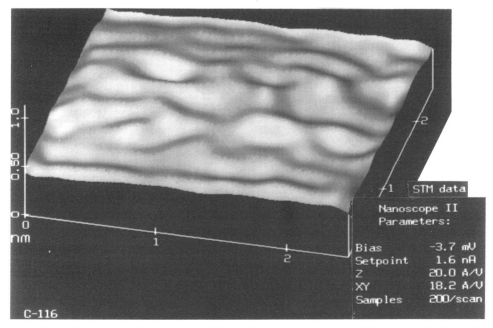

When after "frozen state" interaction of two C_{58} molecules with substrate was "unlocked," molecules rotated and moved. Because there is high affinity of each C_{58} molecule (4 open carbon bonds) and they were very close to one another, two molecules joined together spontaneously, making a C_{116} molecule ("strong dimer").

F:9-7   **SUPERCONDUCTIVITY**
                          C$_{116}$ as C$_{60}$ dimer approach

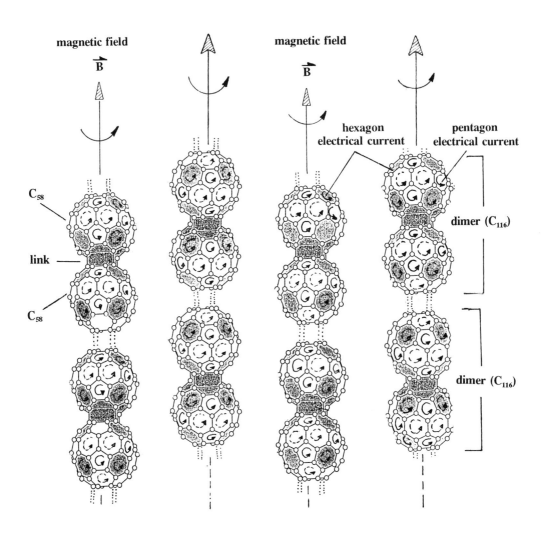

The C$_{60}$ molecule, both in solution and solid state (crystal-thin film), rotates from 10^9 to 3×10^{10} times per second. Pentagon rings produce an electrical current (hexagon rings also have an electrical current, but much less), which produces a magnetic field. Because the C$_{60}$ molecule rotates randomly, the magnetic field is small. If the rotation of the C$_{60}$ molecule could be fixed in only one direction, the magnetic field would be much higher. One possible solution to fix the rotation in one direction is to build a dimer C$_{116}$. In the solid state, dimers will be oriented and produce a magnetic field which can be used together with doping of C$_{116}$ both outside and inside of the molecule.

F:9-8

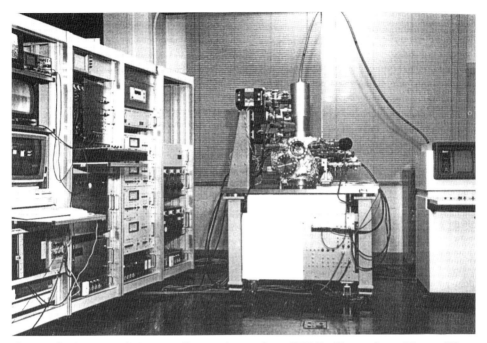

One of the possible configurations for STM (Scanning Tunnelling Microscopy) and high vacuum system for imaging and manipulation on nanometer scale.

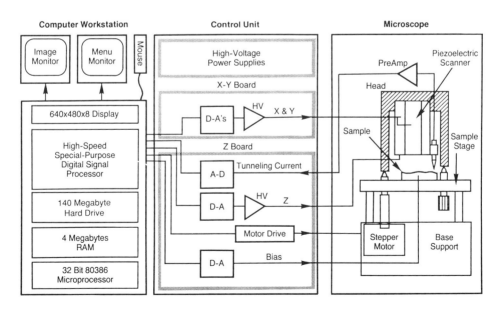

Block Diagram of NanoScope II (Digital Instruments, Inc., Santa Barbara, USA).

CHAPTER 10

ELECTROCHEMISTRY OF FULLERENES
AND ITS APPLICATIONS

10.1 Introduction

Theoretical calculations of the electronic energy levels for C_{60} under the assumptions of perfect truncated icosahedral symmetry and no static Jahn-Telher distortion, suggest that the lowest unoccupied molecular orbital (LUMO) of neutral C_{60} is triply degenerate. (Haddon et al, 1986) Thus, C_{60} is expected to be highly electronegative with the potential of accepting up to six electrons forming anion radicals. The combination of the highly symmetrical structure, which indicate remarkable stability, and of the high molecular orbital degeneracy of C_{60} this molecule could only lead to very rich redox chemistry.

Indeed, extensive work has been reported for the reduction, and intercalation, and even oxidation of modified fullerenes. In general, electrochemical oxidation is expected to be difficult because the ionization potential of C_{60} (Zimmerman et al, 1991) is quite high at 7.50 - 7.72 eV. However, the electrochemical irreversible oxidation of C_{60} and C_{70} has been reported (Dubois and Kadish, 1991) at E_{ox} = +1.76 V vs. SCE at a scan rate of 0.1 V/sec. The results indicate that the oxidation of C_{60} and C_{70} proceed via overall four-electron transfers which are followed or accompanied by one or more chemical reactions to render the overall electrochemical oxidation irreversible. Accordingly, in this chapter we will focus on the electrochemical reduction of fullerenes which should be more favorable.

The electrochemical reduction of fullerenes has been investigated in several different forms including: (1) fullerenes in solution where the fullerenes are either dissolved in non-aqueous solvent with sufficient solubility for the fullerenes such as benzene, toluene, acetonitrile, tetrahydrofuran (THF), etc.; (2) fullerenes thin film on inert conductive substrate, and (3) solid-state fullerenes where the fullerenes are the major constituent of the electrode and polymer electrolyte is used. This classification is by no means inclusive, and several electrochemical arrangements can fall under more than one of these classifications. We will, therefore, identify these cases and describe them once in the first classification and cross reference later classification.

10.2 Electrochemistry of Fullerenes in Solution

In late 1990, just after Krätschmer, Huffman and co-workers (Krätschmer et al, 1990a, and Krätschmer et al, 1990b) reported their findings of a method to produce microscopic quantities of C_{60}, investigations of the electrochemistry of fullerenes began.

Haufler, Smalley and co-workers (Haufler et al, 1990) first reported the generation of ionic forms of C_{60} in solution by cyclic voltammetry. The C_{60} was dissolved in CH_2Cl_2 solution containing 0.05 m of $[(n\text{-}Bu)_4\ N]\ BF_4$ as a supporting electrolyte. The experiments used glassy carbon button electrodes, and platinum wire counter electrode. It was shown that C_{60} undergoes facile electrochemical reduction in two successive steps with $E_{1/2}$ at -0.61 V and -1.0 V relative to NHE. Similar results were reported with platinum working electrodes. This reduction was shown to be reversible. Both of the C_{60} electrochemical reduction reactions were determined by coulombetry to be one-electron processes. The two highly reversible reductions of C_{60} in aphotic solvent suggest that it should be possible to electrochemically produce stable "fulleride" salts. These may result in new materials, and perhaps a new class of rechargeable batteries.

Allemand, Wudl, and co-workers (Allemand et al, 1991) soon after extended Smalley et al.'s work investigating the effect of solvent and the difference between the C_{60} and C_{70} electrochemical behavior. Suprisingly, they found that both C_{60} and C_{70} behave identically. The electron affinity of fullerenes was related to the molecule cage strain energy, and the number of pyracyclene bonds. The larger strain relief would tend to increase C_{70}'s electron affinity relative to C_{60} but the fewer pyracyclene bonds in C_{70} would have the opposite effect. Thus, it is possible that both effects operate and cancel each other out to produce the observed results. Furthermore, Wudl et al, (1991) reported three irreversible reduction waves. The third wave; $E_{1/2}$ = -1.25 V (vs. Ag/AgCl reference) is reversible only at a cycling rate above 1 V/sec. They also reported dramatic solvent (CH_2Cl_2, THF, o-dichlorobenzene ODCB, and benzonitril) effect on the reduction potentials. The first reduction potential was lowered by 0.2 V in THF when compared to the other solvents. The solvent effect was related to the Gutmann solvent donicity number.

Dubois, Kadish and co-workers (Dubois et al, 1990a) subsequently, demonstrated a fourth one-electron reduction for both C_{60} and C_{70}. Cyclic voltammograms of C_{60} and C_{70} in CH_2Cl_2 electrolyte are shown in Figure 10-1. C_{60} exhibits reduction at a scan rate of 20 V/sec at $E_{1/2}$ -0.44, -0.82, -1.25, and -1.72 V vs. SCE. All four reduction processes are reversible at this scan rate, but only the first two appear to be reversible at slow scan rate (0.1 V/sec)

F:10-1   F:10-3

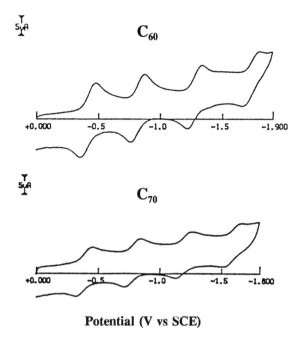

Potential (V vs SCE)

Cyclic voltammograms at a platinum electrode in CH$_2$Cl$_2$, 0.05 M [(n-Bu))$_4$N](BF$_4$), for C$_{60}$ at 10 V/s and C$_{70}$ at 20 V/s.

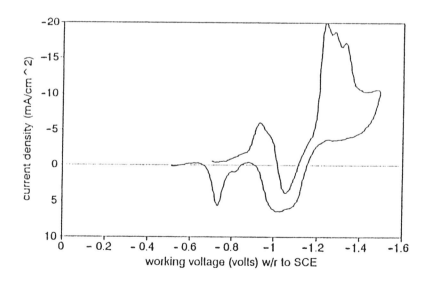

Cyclic Voltammograms for C$_{60}$ Films on Ag Electrodes in 5 N KOH Electrolyte at Scan Rate of 0.025 V/sec controlled-potential coulombetry established that the first two reductions are fully reversible single electron-transfer. Bulk electrolysis to produce C_{60}^{3-}, however, was not reversible. C_{70} also displays four consecutive reductions at $E_{1/2}$ of -0.41, -0.81, -1.20, and -1.58 V vs. SCE. The first three reduction processes were found to be reversible at slow scan rate (0.1 V/sec). This indicates that C_{70}^{3-} is less reactive in CH_2Cl_2 than C_{60}^{3-}. These results support the dominance of molecular strain energy on the electron affinity of C_{70} relative to C_{60}.

Dubois, Kadish and co-workers (Dubois et al, 1991b) further studied the electro-reduction of C_{60} and C_{70} in various solvents to extend the reduction beyond the four electro-reduction. This is typically hindered by the lack of a solvent in which C_{60} and C_{70} are soluble and the solvent has a reduction potential window that extends beyond -2.8 V vs. SCE. In this study, Dubois et al, (Dubois et al, 1991b) utilized benzene, which has one of the widest known reduction windows. Under these conditions, five electron reductions were observed with $E_{1/2}$ of -0.36, -0.83, -1.42, -2.01, and -2.60 vs. SCE. At high scan rate (20 V/sec) all five reductions were reversible. While the third and fourth reductions of C_{60} were irreversible and gave unstable species in CH_3Cl_2[8], these reduction processes produce quite stable fulleride ions in benzene and benzonitrile[9]. Accordingly, the irreversibility of highly electronegative fulleride ions must be due to a follow-up chemical reaction with the solvents. This also indicates the importance of solvents on the reduction potential and degree of reduction.

Xie, Echegoyen and co-workers (Xie et al, 1992) used a mixed solvent system and low temperature to expand the potential window. On the basis of supporting electrolyte and fullerene solubility considerations, the optimal solvent composition was between 15 and 20% by volume acetonitrile in toluene. These conditions have allowed one for the first time to observe the sixth electron reduction for C_{60} and C_{70} leading to the formation of C_{60}^{6-} and C_{70}^{6-}. The highest temperature allowing resolution of the sixth reduction of C_{60} was approximately 5°C.

For C_{70} it was possible to observe the fifth and sixth reduction even at 25°C. It was also interesting to note that under this protocol the multiple reductions are all reversible when observed at low potential scan rate (0.1 V/sec). The cyclic and differential pulse voltammograms of C_{60} and C_{70} at -

10°C are shown in Figure 10-2. The unusual stabilities and clean voltammetric data indicate that it will probably be possible to use these highly electronegative species in electro-synthesis, and as a powerful reducing agent. No practical applications, however, have been reported in the literature. Furthermore, the generation of highly electronegative fullerene species must be carried-out in aprotic solvents in a controlled atmosphere to prevent oxygen and moisture oxidation of the fulleride ions.

Kadish et al, (Boulas et al, 1992) have reported a method to water solubilize C_{60} by means of cyclodextrin inclusion chemistry. Cyclodextrin (CD) can form inclusion complexes where the hydophoblic interior of the cyclodextrin facilies a host-guest interaction with the nonpolar guest, while the highly polar exterior of the CD, rich in hydroxyls, facilies solubilization. This is accomplished by equilabrating carbon disulfide solution nearly saturated with C_{60} with a 30/70 (V/V) water/ethanol mixture containing 3 wt% γ-cyclodextrin. After sonication and nitrogen gas purging for five hours to evaporate the carbon disulfide, the remaining water/ethanol was magenta in color indicating the presence of γ-cyclodextrin/C_{60} compound. Three one-electron transfers were observed by cyclic voltammetry of this solution. All three electron transfer steps are reversible at potential scan rates exceeding 0.2 V/sec with $E_{1/2}$ - occurring at -0.57, -1.08, and -1.46 V vs. SCE. However, only the first couple remains reversible at lower scan rates, while the other two are coupled to chemical reactions. The nature of the follow-up chemical reactions were not identified. The practical application of solubilized C_{60} in aqueous solution, given the fact and only the first electron transfer will be stable, is not clear.

In summary, fullerenes in aprotic solvent can be reduced to hexanionic molecule confirming theoretical predictions (Hadon et al, 1986) that C_{60} should be able to accept six electrons to form diamagnetic C_{60}^{6-}. The selection of solvent system and operating temperature allows for the formation of stable and reversible C_{60}^{n-} and C_{70}^{n-}. These highly electronegative species should find application in electro-synthesis, and possibly in homogenous bulk reduction of difficult to reduce compounds.

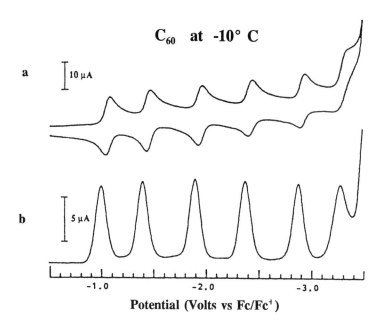

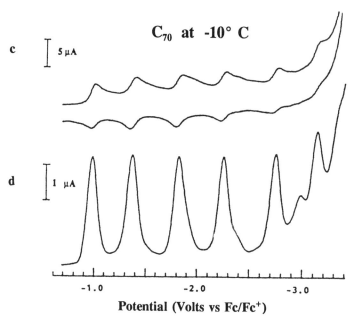

Reduction of C_{60} and C_{70} in CH_3CN/toluene at $-10°C$ using (a and c) cyclic voltammetry at a 100 mV/s scan rate and (band d) differential pulse voltammetry (50-mV pulse, 50-ms pulse width, 300-ms period, 25 mV/s scan rate).

10.3 Electrochemistry of Fullerenes Thin Films

Bard et al, (Jehoulet et al, 1991) described the first electrochemical study of films of C_{60} on platinum electrodes. The films were prepared by evaporation on the electrode surface of a few microliters of a solution of C_{60} in CH_2Cl_2 or benzene. The reduction of C_{60} film in acetonitrile showed four reduction peaks with cathodic peak potential (Epc) at -0.71, -0.93, -1.42, and -1.78 V vs. SCE. The first three peak potentials are close to those reported for dissolved C_{60} (Haufler et al, 1990 and Allemand et al, 1991), and probably represent successive one-electron transfer reactions. However, the behavior on scan reversal is different than that of dissolved C_{60} since the anodic peaks are displaced to potentials positive relative to the cathodic peaks. Furthermore, unusual hysteresis was observed, and continuous scanning showed slow and continuous decrease in peak currents. This behavior was attributed to supporting electrolyte co-intercalation (Zhou et al, 1992) (tetra-butylammonium, TBA^+ ions) possible increased solubility of ionized forms of C_{60} in the solvent. The large splitting between the waves implied significant energies of reorganization of C_{60} film as results of TBA^+ ions co-intercalation.

Similar behavior (Segar et al, 1991) was observed for C_{60} films electro-reduction in propylene carbonate/lithium perchlorate electrolyte that is commonly used in lithium batteries. Cyclic voltammetry showed that reduced C_{60} and C_{70} are partially soluble in the electrolyte, which results in poor electrochemical reversibility.

Bard et al (Hehoulet et al, 1992) further studied the electro-reduction of C_{60} and C_{70} films showing that upon reduction of fullerene films in acetonitrile, TBA^+ or larger cations can penetrate into the crystallitesm, causing a large change in crystallite structure. The film can be unoxidized to recover the C_{60} or C_{70} structure. Although electrochemical transformation of fullerenes and quaternary ammonium fulleride films is possible, neither are good electronic conductors. However, the conductivity and porosity of the films appears to be sufficient for almost complete reduction and re-oxidation at slow scan rates. The nature of the solution cation is important in the redox process, and different behavior is found with K^+ and Cs^+. In the latter case, a more reversible intercalation of the cation is suggested, but the reduced forms $M^+C_{60}^-$ appear to be much less stable than $TBA^+ C_{60}^-$ films.

Accordingly, electrochemistry of fullerenes films in aprotic solvents with possible limited solubility in solvent and intercalation of the cation of the supporting electrolyte is complex and offer no significant potential for practical applications because of its irreversibility and hysteresis.

At MER we investigated the electrochemistry of C_{60} films on Ag electrodes in aqueous electrolyte for the reduction of hydrogen. The C_{60} films were prepared by vacuum vapor deposition of C_{60}. A dense film of 0.5 - 2.0 μm was formed. The electrolyte was 5.0 M KOH to simulate Ni/Cd battery commercial electrolyte. In this electrolyte it is expected that C_{60} will have no solubility; accordingly, the complication of solvent-supporting electrolyte co-interaction should be absent.

Cyclic voltammetry of C_{60} films in aqueous electrolyte is shown in Figure 10-3. This is the first electrochemical reduction of C_{60} film electrolytes in aqueous electrolyte to be reported. Two reduction potential peaks were observed at -0.90 and -1.26 V vs. SCE. The current for first reduction is proportionally smaller than the second reduction. Furthermore, the current becomes positive after the first reduction indicating the formation of strangely absorbed species. However, the current associated with the second reduction is significantly larger than that of the first reduction process. The anodic peaks are displaced to potential only about 0.2V of the cathodic peaks. The scan rate for this cyclic voltammetry was very slow at 0.025 V/sec; accordingly, higher degree of reversibility is expected at higher scan rates. It is interesting to note that there was no gas evolution up to -1.5 V vs. SCE; therefore, these reduction processes are associated with the C_{60} film. Coulometric measurement at a potential after the second reduction process (-1.3 V vs. SCE) was performed on a geometrically well-defined C_{60} film electrode, and results showed the electrochemical reduction was fully reversible, at least for the first two cycles. Further cycling resulted in film spalling. Improved film adhesion will be required for developing practical applications. The number of charge transfer per C_{60} molecule was calculated from the coulometric results and they are summarized in Table 10-1.

It is interesting to note that the charges transformed per C_{60} molecule is a very large and even number. Since C_{60} can only accept six electrons, the observed charge transfer of C_{60} film electrode in aqueous electrolyte must be associated with chemical reaction. In this case, the most likely reaction is

Electrode Diameter (cm)	0.3175
C_{60} Film Thickness	2×10^{-4}
Volume of C_{60} (cm^3)	1.58×10^{-5}
Weight of C_{60} (gm)	2.7×10^{-5}
Moles of C_{60}	4.51×10^{-7}
Charge Capacity (mAhr)	0.68
Ratio of Charge/C_{60}	56

Table 10-1:   The Degree of Electrochemical Reduction of C_{60}

hydrogenation. If this assumption is correct, then the charge of C_{60} films must have produced $C_{60}H_{56}$. Laser ablation mass spectroscopy of a sample of the charged C_{60} film is shown in Figure 10-4, showing a mass of 770, which is consistent with a formula $C_{60}H_{56}$. It is also interesting to note that there was no signal at 720, corresponding to C_{60}, indicating full hydrogenation of the C_{60} film. These results are significant, providing the hydrogenated C_{60} compounds are stable, in developing practical applications of C_{60}. Previous work (Haufler et al, 1990 and Jehoulet et al, 1992) has demonstrated the hydrogenation of C_{60} by Birch reduction (Haufler et al, 1990) to $C_{60}H_{36}$ and by gas phase catalytic technique (Jehoulet et al, 1992) yielding $C_{60}H_{36}$ and $C_{70}H_{46}$. High pressure (400 psi) catalytic hydrogenation of C_{60} in ethylacetate solvent with platinum oxide catalysts was also performed by MER and $C_{60}H_{60}$ was produced. Accordingly, this is the first time that electrochemical hydrogenation has been reported. However, the resultsare in agreement with the previous work, in terms of the degree of hydrogenation.

10.4 Electrochemistry of Solid-state Fullerenes

Chabre et al, (1992) have succeeded to intercalate C_{60} with lithium using all solid-state electrochemical cells and polymer electrolytes. This offered an

F:10-4

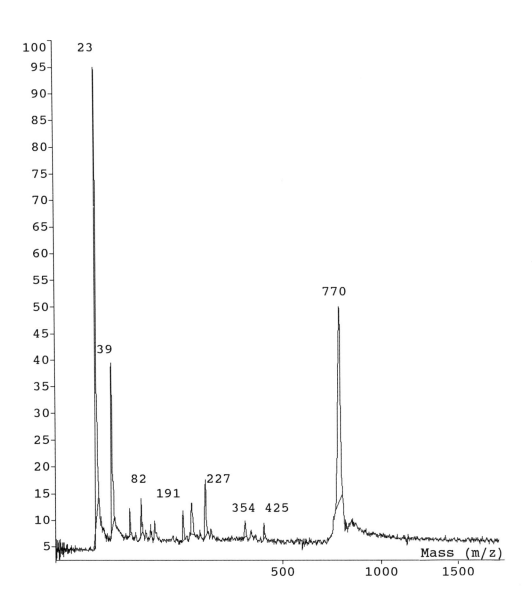

Mass Spectragram of charged C_{60}-Ag Electrode Sample (C_{60} H_{56})

advantage over previously reported work on electrochemistry of fullerenes in aprotic solvents, and thin films in aprotic electrolytes. Polymer electrolytes generally exhibit larger electrochemical stability windows than liquid electrolytes. In this work, a composite electrode of 60% pure C_{60} and 40% solid polymer electrolyte ($P(EO)_8$ $LiClO_4$) were cast from acetonitrile suspension into a stainless steel current collector. These composite electrodes were used in a coin type cell with metallic lithium electrode, and polymer electrolyte film. The cells were operated at 80°C to maintain the electrolyte in the high ionic conductivity amorphous phase. Three well defined reduction peaks were observed at 2.3, 1.9 and 1.5 V vs. Li reference. A fourth peak was observed as a shoulder at 1.0 V on very large fifth wave at 0.8 V vs. Li. By integrating the current vs. time, x in Li_xC_{60} was determined for the successive electron plus in injection processes and x was found to correspond to 0.5, 2, 3, 4 and 12. Note that the Li:C ratio is about the same for $Li_{12}C_{60}$ and LiC_6, the saturated value for the Li intercalated graphite compound. However, reduction beyond the third wave is irreversible and involves extensive reorganization of the C_{60} solid. Accordingly, the viable Li:C ratio is Li_3C_{60} or LiC_{20}. For lithium battery application this represents poor electrode capacity.

Chabre (Chabre et al, 1992) also studied the electrochemical intercalation of solid C_{60} electrode with sodium in the same technique as described above except they used $P(EO)_n$ $NaCF_3SO_3$ solid polymer electrolyte. They reported a rather rapid first intercalation/reduction in the form of voltammograms. Five successive reductions in all were observed similar to the Li/C_{60} system at about the same potential positions and with similar kinetics behavior. The two first steps at 2-3 V and 1.95 V were attributed to the formation of new phases $Na_{0.5}C_{60}$ and Na_2C_{60}. The third step forms Na_3C_{60} in homogenous solution process from Na_2C_{60}. AFter the second reduction, however, self-discharge was observed. That was attributed to a slight solubility of reduced C_{60}^{n-} species in the Na^+ conducting P(EO) based electrolyte.

Accordingly, while the solid state electrochemical intercalation of C_{60} is interesting, it does not offer significant advantage over standard graphite

in term of specific capacity for battery application. It, however, could offer a means for producing doped fullerene compounds.

10.5 Electrochemical Hydrogenation of Fullerenes

The feasibility of hydrogenation of C_{60} by chemical (Compton et al, 1992) (Birch reduction), catalytic (Chatre et al, 1992) and electrochemical techniques have been demonstrated. MER Co. (Dr. Loutfy) investigated the practical application of electrochemical charging and discharging of C_{60} solid electrode in aqueous electrolyte. C_{60} electrodes were prepared by mixing C_{60} with conductors with high hydrogen overvoltage such as Ag and uniaxial pressing at relatively high pressure to produce near full density electrodes. Cyclic voltammogram of C_{60}-17 vol% Ag is shown in Figure 10-5.

A negative current peak appears at -1.6 V and reversed current peak appears at -0.7 V vs SCE. These peaks are certainly associated with hydrogen evolution and C_{60}. In a second experiment, the voltage was fixed at -1.6 volts and significant current was observed without gas evolution, which supports the hydrogenation of C_{60} concept.

The electrochemical behavior of C_{60}-15 vol% Ag electrode under high hydrogen pressure was also investigated. The high hydrogen pressure cyclic voltammetric experiments were carried out in the high pressure reactor.

Cyclic voltammograms for C_{60}-15 vol% Ag electrodes were also investigated under atmospheric and high hydrogen pressure (175 psi) and those results are shown in Figure 10-6. The key difference in addition to the increased current is the well defined charging peak for the hydrogenation of C_{60} under high hydrogen pressure. This again establishes the feasibility of the proposed hydrogenation of C_{60} concept. The implication of this result is that hydrogenation of C_{60} under high pressure could result in higher specific energy for the proposed system.

Electrodes of C_{60}-Ag were also prepared for charge-discharge characterization. Typical charge-discharge characteristics for the C_{60}-Ag electrode at 10mA/cm^2 are shown in Figure 10-7. It is interesting to note that during charging the voltage quickly dropped from the OCV of typically -0.35 V to < - 1.0 V within the first few minutes. The charging voltage seems to

F:10-6

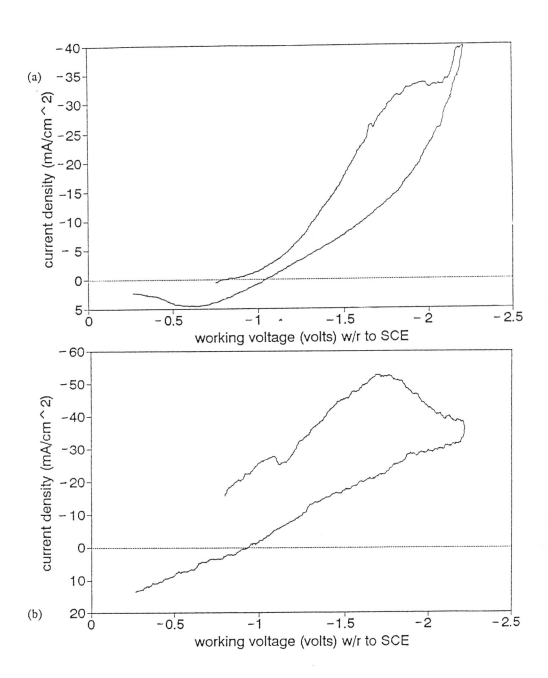

Cyclic Voltammograms of C_{60} - 15% vol Ag in 30% KOH (a) Under Atmospheric Pressure, (b) Under 175 psi H_2 at 0.25 mV/sec Scan Rate.

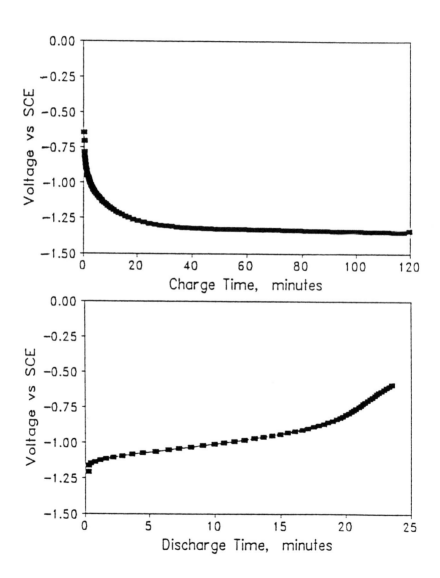

Typical Charge-Discharge Behavior of C_{60}-Ag Electrode in 30% K
Electrolyte at 10 mA/cm^2.

level off at -1.3 V for a significant duration without any noticeable gas evolution. During discharge the voltage instantaneously changes from -1.25 V to -1.20 V and slowly decays. After about 20 minutes there appears to be voltage deflection. In this discharge period the nominal voltage is about -1.0 V. This compares very favorably to the nominal discharge voltages for the Cd electrode of -0.95 V.

The effect of charge and discharge current density on the performance of the C_{60}-Ag electrodes was also evaluated by repeat charge and discharge cycling. The results of charging at 5, 10 and 20 mA/cm^2 are shown in Figure 10-8. As can be seen, there is an increase in voltage from nominally -1.10, -1.20, and -1.30 V, respectively. The voltage appears to be leveling at -1.30 V, which corresponds to the hydrogen evolution potential. The results of discharging at 10 and 20 mA/cm^2 and 10, 5, and 2.5 mA/cm^2 are shown in Figures 10-9 and 10-10, respectively. As can be seen, the discharge voltage is quite sensitive to current density, particularly at the high current density. However, at or less than 10 mA/cm^2 discharge current the nominal discharge voltage is $\leq$ -1.0 V which is equal or better than the Cd electrode.

Particularly at a discharge current of 2.5 mA/cm^2, the C_{60}-Ag electrode is exceptionally better than the Cd electrode by about 200 mV. The relatively large effect of current density on the discharge voltage of C_{60}-Ag electrode can to some degree be anticipated because of the nature of the chemical reaction involving bond formation between the C_{60} molecule and the hydrogen. The current density limitation of the C_{60} electrode can be easily overcome because C_{60} can be applied from vapor or liquid to form thin films and thus producing high surface area electrodes.

The capacity of the C_{60} electrode during charging and discharging was estimated from the voltage-time curve, and the current density. The results are summarized in Table 10-2.

It is clear that the capacity both for charging and discharging decreases with current density. However, excellent charge recovery or efficiency of $\geq$ 92% was obtained at $\leq$ 5 mA/cm^2 current density.

F:10-5   F:10-8

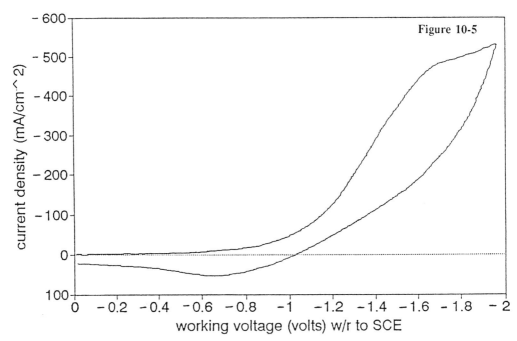

Cyclic Voltammogram of 85% C_{60}/15% Ag Electrode in 30% KOH Electrolyte
at 20 mV/sec Scan Rate

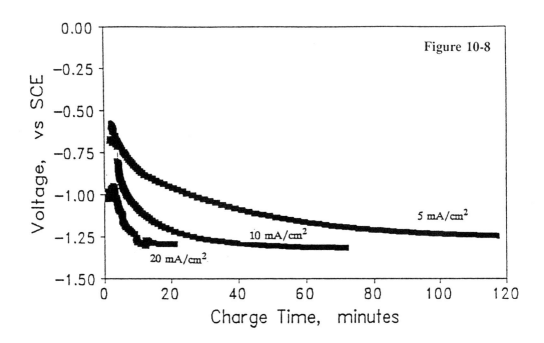

Charge of C_{60}-Ag Electrodes at Various Current Densities

F:10-9 **F:10-10**

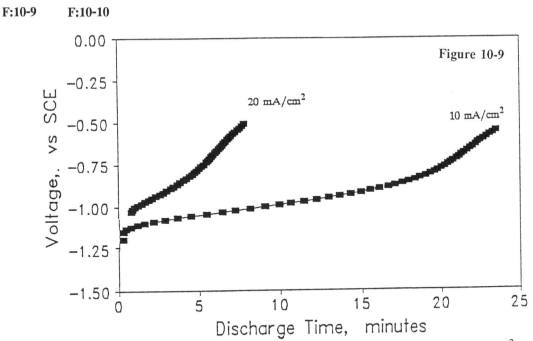

Discharge of C_{60}-Ag Electrodes in 30% KOH Electrolyte at 20 and 10 mA/cm².

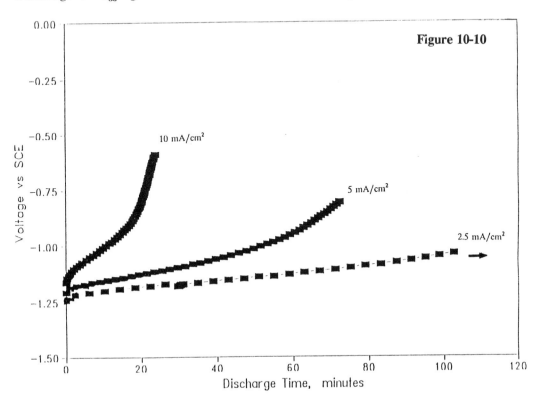

Discharge of C_{60}-Ag Electrode in 30% KOH, at 10, 5, and 2.5 mA/cm²

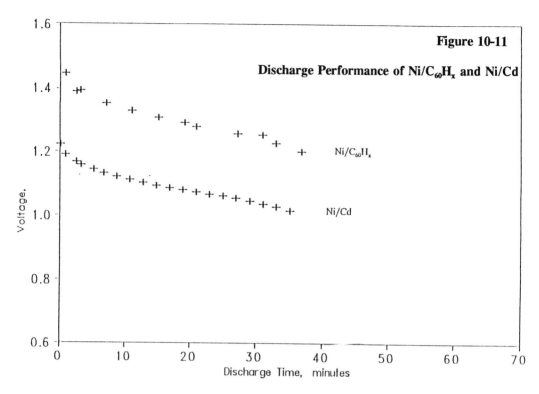

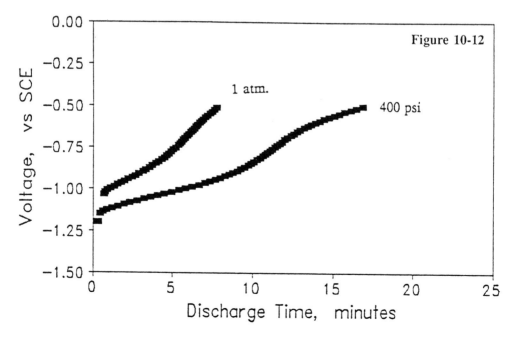

Discharge Behavior of C_{60}-Ag Electrode in 30% KOH at 20 mA/cm^2

The discharge performance of $C_{60}H_x$ electrode at 2.5 mA/cm^2 discharge current density vs. the commercial nickel oxide electrode is

Current Density	Charge Capacity Ahr/cm^2	Discharge Capacity Ahr/cm^2	Efficiency
2.5	-	0.0046 *	-
5	0.0063	0.0058	92
10	0.0058	0.0042	72
20	0.0050	0.0030	60

Table 10-2. Charge and Discharge Capacity of C_{60}-Ag Electrodes (* Did not completely discharge)

calculated from the voltage difference between the two electrodes and is plotted in Figure 10-11. In addition, the discharge performance of NiO/Cd electrodes was calculated and is shown in Figure 10-11 for comparison. Even though the discharge current density for the $C_{60}H_x$ is low, the performance of this new cell is exceptional in performance with OCV of about -1.5V and nominal discharge voltage of -1.3 V. These results demonstrate the feasibility of the proposed concept in electrochemically hydrogenating and dehydrogenating C_{60} and also in demonstrating that the $C_{60}H_x$ electrode is a viable electrode to replace Cd in Ni/Cd batteries.

The charge-discharge characteristics of C_{60}-Ag electrodes were also evaluated under high hydrogen pressure. This could simulate operating under sealed battery conditions. The discharge behavior of C_{60}-Ag electrode under 400 psi hydrogen at 20 mA/cm^2 current density is shown in Figure 10-12. This result is also compared to discharge under atmospheric conditions. It is very interesting to note that improved discharge performance is obtained when discharge is carried out under high pressure even at a high discharge current density of 20mA/cm^2. This result is very significant in that it overcomes the

C_{60} FOR FUEL CELL APPLICATION

Charge:

$C_{60} + xH_2O + xe$	$= C_{60}Hx + xOH^-$	Cathode
xOH^-	$= x/2\ H_2O + x/4\ O_2 + xe$	Anode
$C_{60} + x/2\ H_2O$	$= C_{60}Hx + x/4\ O_2$	*Net Charge*

Discharge:

$C_{60}Hx + xOH^-$	$= xH_2O + xe$	Anode
$x/2\ H_2O + x/4\ O_2\ \text{(air)} + xe$	$= xOH^-$	Cathode
$C_{60}Hx + x/4\ O_2\ \text{(air)}$	$= x/2\ H_2O$	*Net Discharge*

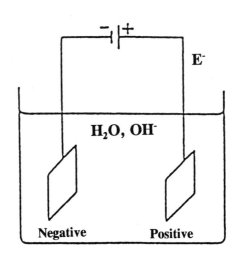

limitations of $C_{60}H_x$ electrode operation at low current density.

In summary, C_{60} electrodes can be electrochemically charged with hydrogen, and the charged $C_{60}H_x$ electrode can be discharged with high efficiency at relatively high current density and with nominal voltage of equal or better than -1.0 V vs. SCE. Silver in C_{60}-Ag electrodes, while adequate in providing the conductivity and the high hydrogen overvoltage, is not an economical alternative to be used in commercial battery applications. Accordingly, alternative conductors are under investigation.

The degree of hydrogenation of mixed C_{60}-Ag was also investigated using non-destructive techniques. To determine C_{60} electrode capacity, a sample of the charged C_{60}-17% Ag electrode was washed, vacuum dried, analyzed using Elastic Recoil Detection (ERD) Spectrography of hydrogen and carbon determination. In this technique, α particles are bombarded on the sample surface with certain angles and the emitted protons are detected after filtering the α particles. This technique can detect hydrogen at concentration levels of 0.1 atomic percent and surface coverage of less than a monolayer.

The results of ERD runs indicate that very little hydrogen was present in the first micron of thickness of the electrode. A very high potassium count was observed instead. The electrode surface was scraped lightly to expose a fresh surface of the electrode and an ERD run was performed. Significant increase in hydrogen was detected with no apparent reduction with time and beam exposure. This without doubt establishes the nature of the electrochemical reduction of C_{60} in aqueous electrolyte to hydrogenation reaction. The composition of hydrogen in C_{60} was estimated to be $C_{60}H_{18}$ in a 1.0 to 1.5μm thick layer. The fact that the hydrogen was uniform through this thickness could indicate that hydrogen penetrated deeper into the pressed electrode. We do not know exactly the thickness of the material scraped off and whether this layer has higher H/C_{60} ratio.

Based on these results, we calculated the capacities of $C_{60}H_x$ electrodes was estimated to be 650 mAhr/gm. This electrode capacity is expected to increase by operating under pressure and at low current densities.

As can be seen, the capacity of $C_{60}H_{18}$ electrodes is still excellent compared to the best metal hydride electrode (which is about 320 mAhr/gm).

ELECTRONIC STRUCTURE OF DOPED C_{60}

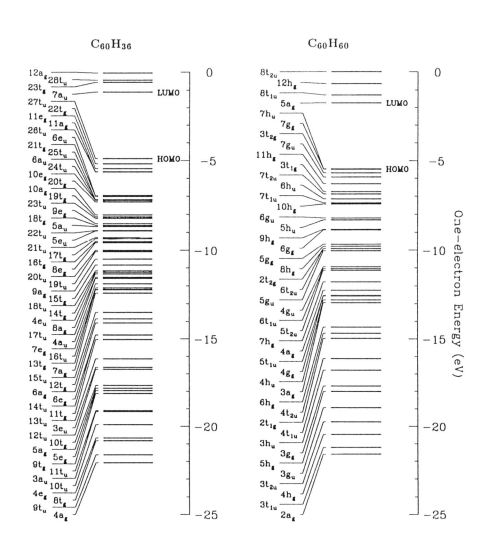

The one-electron eigenvalues of $C_{60}H_{36}$ and $C_{60}H_{60}$ as an approximation based on one-electron eigenvalue. This approach works very well and has been tested for C_{60}, C_{80}, and C_{240}. The energy gap (HOMO-LUMO) of $C_{60}H_{36}$ is about 3.80 eV, while that for $C_{60}H_{60}$ is about 3.60 eV.

10.6 Applications of Fullerene Hydrides

10.6.1 Introduction

Several methods, including Birch reduction and catalytic hydrogenation, have demonstrated the possibility to synthesize hydrogenated compounds (Haufler RE, Conccino J, et. al., 1990),(Petrie S, Javahery G, Wang J, Bohme DK, 1992) of the structure C_nH_x, where 1 x   n. MER Co's research has overcome the complexity and very low process yields of these methods by employing an electrochemical method to hydrogenate fullerenes to levels approaching x = n, as was discussed in the previous section.

This is accomplished by placing fullerenes in intimate contact with conductive materials to form an electrode, which may be employed as the negative electrode in an electrochemical cell, analogously to the negative electrode in a nickel-metal hydride battery. Because of the much higher density of hydrogen storage per gram of fullerene which is theoretically possible than in the case of typical metal-hydride materials, fullerenes offer promise to extend the useful applications of metal-hydrides which have already been established.

Such applications fall broadly into two categories: (1) those based on the ability to collect and store hydrogen safely and in compact form; and (2) those based on the electrochemical and thermodynamic consequences of such storage. One of the most promising applications for the electrochemical hydrogenation of C_{60} is as an alternative for the metal hydrides in a nickel-metal hydride battery for storage of electrical energy.

10.6.2 Reversible storage of hydrogen gas

One may use materials C_nH_x to store hydrogen compactly for use in hydrogen purification or gettering, hydrogen combustion engines, or for hydrogen-air fuel cells. The solid may be stored until the hydrogen gas is required, at which time the hydrogen gas may be liberated electrochemically, or by chemical dehydrogenation. This also opens the door for new methods of hydrogen compression and purification, heat pumping, and refrigeration.

POSSIBLE APPLICATIONS

OF

C_{60}

Physical Properties

Lubricants   High Strength Fibers

Molecular Membranes

CVD Films, Diamonds   Abrasives

Acoustical Sensors   Hydrogen Storage

Fuels Catalysts

Organic Solvents

Semiconductors

Non-Linear Optical Devices   Medical Imaging Agents

High Energy Batteries

Superconductors

Chemical Sensors   Ethical Pharmaceuticals

Electro-optical molecular devices

Electrical & Optical Properties

Chemical Properties

10.6.3 Hydrogen storage in hydrocarbons

Experimentally, C_nH_x may be used as a clean- and rapid-burning hydrocarbon fuel. In these applications, the preferred electrode for hydrogenation is a film of C_n applied to a conductive foil or film from which the product may be easily removed by mechanical means such as scraping or flexing the substrate after charging, although care must be taken to ensure that accidental spontaneous combustion does not occur.

10.6.4 Rechargeable (secondary) electrochemical cells

Such cells may be used either for storage of hydrogen for uses listed above, or they may constitute the elements of a battery for storage of electrical power. The author has demonstrated that fullerene hydrides may be used to replace the cadmium electrode in a Ni-Cd battery.

10.6.5 Primary batteries

In addition to fully rechargeable batteries, one may produce electrode materials C_nH_x, where x is nearly or identically equal to n, and thus holds the maximum amount of electrochemical energy possible using this system. Batteries with electrodes manufactured in this way might not be fully rechargeable, but would offer exceptionally high energy density, as indicated in the table below:

Composition	Specific Capacity, mA-hr/gm
$C_{60}H_{60}$	2250
$C_{60}H_{36}$	1340

Theoretical Capacity of $C_{60}H_x$ Electrodes

Birch Reduction of C_{60}

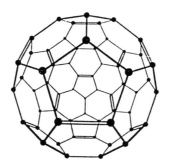

Hydrogenation $C_{60}H_{36}$

One possible isomer of $C_{60}H_{36}$ (the hydrogens are not shown), where the remaining 12 double bonds are isolated from each other, one on each pentagon.

1.   156 mg C_{60}, suspended in 100 ml liquid NH_3

2.   Add 15 ml t-BuOH

3.   Add 365 mg Li [240 molar equivalents = $4e^-$ equivalents]

4.   Stir in reflux until solution is discharged

5.   Black particles may still be evident
     Add another 220 mg Li and 5 ml t-BuOH

6.   Solution turns blue, reflux until color is discharged

7.   Color becomes light beige

8.   Evaporate NH_3, suspend in H_2O

9.   Extract with toluene

10. Evaporate toluene extracts, vacuum dro
     Yield = 163 mg off-white solid

RF value = 0.37 using 30% methylenechloride in hexane
Not UV reactive
Revealed w/IV

These values compare quite favorably with the best metal hydrides known. The efficiency with which experimental cells store and deliver energy is also extremely attractive, reaching values as high as 92% when operated at low current densities. Comparable maximum efficiency of sealed Ni-Cd cells is about 71%, while vented Ni-Cd cell maximum efficiency ranges from 67% to 80%.

10.6.6 Conclusions

In summary, fullerene hydrides offer a rich variety of useful applications, for hydrogen storage as an end in itself, for thermodynamic applications in cryocoolers and heat pumps, and for use in batteries and hydrogen purification applications. Experimental data implies that fullerene hydride batteries may soon offer a non-toxic, lightweight, highly efficient rechargeable class of batteries. Assuming cost barriers in the production of starting materials can be overcome, we may also even have fullerene hydride solid fuels, all resulting from a single discovery, that of an electrochemical method to efficiently produce fullerene hydrides.

10.7 Practical Applications of Fullerenes

In the years since Huffman and Krätschmer's discovery of the method to produce gram quantities of solid fullerenes, research has blossomed in many areas of commercial and scientific importance, beyond the narrow applications of electrochemistry described earlier in this chapter.

The unusual structure of these molecules has been compared to a 3-dimensional benzene ring. Like benzene, buckminsterfullerene forms the basis for a whole branch of chemistry. Already, we know that C$_{60}$ is not the only molecule in this family of materials. Experiments have demonstrated the formation of related molecules ranging in size from 28 to many hundreds of carbon atoms, all sharing the same basic spheroidal structure. Recently, researchers at NEC showed that one may employ conditions nearly identical to those required to make fullerenes, as this family of compounds is known, to synthesize tubular molecules best described as sheets of graphite rolled into cylinders with half spherical caps on each end. The new nanotubules are called buckytubes, just as the fullerenes are called buckyballs.

In this section, we address what factors affect the further development, introduction and adoption of those products. In short, what are fullerenes good for, and why?

By examining the unusual properties (See Figure 15) exhibited by fullerenes one begins to understand why corporate researchers are so enthusiastic. Proposed commercial applications include abrasives, lubricants, superconductors, photoconductors, precursors to thin film diamond synthesis, constituents in so-called "super alloys," high energy batteries, molecules for hydrogen storage, medical imaging and therapeutic agents, catalysts, semiconductors, sensors, rocket propellants, nonlinear optical devices, ferromagnetic complexes, carriers of radioactive isotopes, high strength microfibers, semi-permeable membranes for the separation of mixed gases, and a whole new generation of fullerene polymers.

The descriptions which follow provide an overview of several of the applications that appear to hold the greatest promise for near- and medium-term products derived from fullerene compounds. One should note, that none of these possible applications has yet evolved into a commercially available product. On the other hand, this analysis does not address the many products for which no current market can be defined, which almost certainly will develop from the R&D now in progress. The goal of this section is to describe briefly several (but by no means all) of the emerging applications for fullerenes, and the evidence which supports the practical use of fullerenes within them.

Research has proven experimentally viable applications for fullerenes in each of these fields, though all will require some additional product development to bring cost and performance into a range competitive with today's technologies. The real excitement in this business lies in those applications which only now are beginning to emerge as a consequence of the unique structural and electrochemical properties of buckyballs.

Fullerenes are still quite expensive for pure carbon, ranging in price from a few tens of dollars to several thousands of dollars per gram. Therefore, a conclusion one may draw about the near term evolution of practical uses for fullerenes is that the uses must provide a high value (compared to other carbon applications) in order to overcome today's relatively high unit cost. This requirement will become less limiting over time, as the price drops.

10.7.1 Fullerene-doped Photoconductors

At the heart of every plain paper copy machine, facsimile machine, and laser printer sold is a specially treated plate, coated with a photoconductive material. These materials are sensitive to light, and they act as the medium upon which the image to be produced is stored. At the beginning of a cycle, the photoconductor plate is given a uniform electrical charge, evenly distributed over the surface. The plate remains in this state until light strikes the surface. For those regions where the light is above a certain threshold intensity, the material changes from being an electrical insulator to being an electrical conductor, and the charge on the surface is allowed to flow off, leaving a map of charged and uncharged areas on the plate which correspond to the dark and bright regions in the desired image.

Older photoconductive materials in use depend on selenium to act as the sensitizer, which makes the coating on the imaging drum photoconductive. More advanced organic photoconductive polymers are displacing the selenium based materials already. DuPont researchers, however, recently disclosed that by doping a PVK polymer with as little as 1% C_{60} (it may be a mixture of C_{60} and C_{70}) by mass, they can create a whole new class of high performance photoconductors. Xerox is reportedly working on a similar product. DuPont has filed several basic patents on this application for fullerenes.

The properties which make this new photoconductor desirable are that the image resolution is as good or better than competitive materials, and that the durability may be much better than selenium-based products. Its performance is in fact "comparable to some of the best commercial photoconductors such as the thiapyrylium dye aggregates."

10.7.2 CVD Diamond Films

Fullerenes, being pure carbon molecules with geometries defined by the number of atoms in each buckyball, may have an important role to play as materials to provide uniform nucleation sites for diamond thin film growth. One of the unique properties of fullerite materials is the low temperatures at which they sublime (for C$_{60}$, this is about 600°C). This makes vapor coating of irregularly shaped surfaces with fullerite relatively straightforward. Furthermore, because fullerenes are soluble in polar solvents such as benzene or toluene, liquid preparations may be applied directly to complex surfaces in a room temperature process, leaving a film of buckyballs behind after the solvent evaporates.

In a recent disclosure by a team of Northwestern University researchers led by R. P. H. Chang, investigators claim they've discovered simple way to use fullerenes to crystallize thin films of diamond (Wang et al, 1992). "The Northwestern team's recipe is based on a buckyball variant with 70 carbon atoms. After placing a thin layer of the balls on a silicon surface, the scientists blast them with electrically charged particles -- a step that apparently rips open the tops of the balls to form molecular structures conducive to diamond formation. Using CVD methods, the scientists then passed a mixture of natural gas and hydrogen over the frayed buckyballs. The balls act as seed crystals, allowing many tiny diamonds to form from the carbon gas.

"Using a variation on this theme, researchers may be able to form diamond film resembling a large, single crystal for use in electronic applications, Dr. Chang added."

If this preliminary result is correct, this would be the most important advance in CVD diamond methods ever, allowing realization of the dream of single crystal diamonds, made to order. In polycrystalline applications, the application of C$_{70}$ increases diamond formation on a silicon substrate by 10 orders of magnitude (Buckyballs, 1991).

Applications for diamond thin films include many military areas, such as impact-resistant armor coating, optical / X-ray / particle-beam windows, semiconductor wafers, hard-surfaced gear teeth, diamond-fiber reinforced composites, temperature and radiation hardened electronic components. Other commercial applications arise from properties listed in the table below.

As a semiconductor material, diamond has properties that make it far superior to today's silicon. Japan has been investing R&D dollars at about 10 times the US rate of $10 million annually because of the advantages that are evident in semiconductor applications of this material which is "superbly strong, light, with dielectric properties resembling a perfect vacuum, bulletproof, better at conducting heat than a silver spoon, insoluble in boiling acid, radiation-hard and non-toxic." Today's revolution in the synthesis of thin film diamond promises to make such films relatively inexpensive, with ultimate bulk costs of less than a dollar per gram.

10.7.3 Catalysts

Catalysts today are important in three broad application markets: petroleum refining, chemical processing, and emissions control. Fullerenes offer significant promise as the basis for a new class of catalysts. Scientists including Dr. R. Smalley have proposed that one might, by filling the hollow space inside a fullerene molecule with a metal atom known to have useful catalytic properties, such as platinum or palladium, create a catalyst where the active sites are "protected" by the carbon cage.

Another type of fullerene-based catalyst was recently demonstrated at the Toyohasi University of Technology (Toyohashi University, 1992). A group led by Nagashima "used C_{60} to produce a palladium-carrying polymer complex with high catalytic activity." The polymer is synthesized in a room-temperature reaction in a benzene solution of C_{60} mixed with a palladium complex. An insoluble polymer precipitate forms, whose composition may be adjusted by changing the reaction conditions. Structurally, the polymer is quite unique. It consists of a three-dimensional cross-linked arrangement with each C_{60} molecule coordinated by six palladium atoms.
The DD:S&T article goes on to state that "the polymer complex, which is reportedly the first catalyst carrier ever developed that has a regular shape at the molecular level, was found to catalyze hydrogenation of diphenyl acetylene at normal temperature and pressure.

"The Toyohashi discovery is significant in that it is the first time a cluster of just a few atoms of a material has been shown to catalyze a reaction. Generally, catalysts can function only with much greater mass."

This announcement is notable not only for its immediate implications in the market for hydrogenation catalysts, but also because the catalyst carrier polymer complex can almost certainly be applied to other types of atoms than palladium, and to catalysis applications other than hydrogenation. If the very high catalytic action of this compound is typical of other similar fullerene-based catalysts, there are sure to be applications where high efficiency and low total catalyst mass or volume are important

10.7.4 Abrasives (Including Industrial Diamonds)

Interestingly, fullerenes have been proposed to act not only as lubricants, but also as a new class of super-hard abrasives. On a molecular level the individual C_{60} spheres are unusually hard. Their geodesic structure makes them particularly resistant to compressive forces. This fact leads one to conclude that fullerite should be an excellent abrasive. However, the weak intermolecular bonds between C_{60} molecules result in a low shear strength for the pure material (it was this characteristic that prompted some to suggest using fullerenes as lubricants). Direct application of fullerenes as abrasives are as yet unproved.

However, a very promising approach for using fullerenes as abrasives is to *convert fullerenes to diamond directly, by the application of high pressure at room temperature*. Researchers Regueiro, Monceau, and Hodeau in Europe have disclosed the ability to crush refined mixed fullerenes into diamond at room temperature (Reguiero et al, 1992). The pressure employed was high, about 25 GPa, and the reaction progressed rapidly. This suggests that a practical technique might employ explosive or other shock wave compression of bulk refined fullerenes for low cost production of industrial grade diamonds.

10.7.5 Superconductivity

One surprising result of lab research in fullerenes is that when fullerite is doped with alkali metals such as potassium and rubidium, it becomes a "warm" superconductor. According to A. F. Hebard of AT&T's Bell Labs, K_3C_{60} has transition temperature, T_c, of 18K, while Rb_3C_{60} makes the transition at 28K (Teresco, 1991). More recent reports have increased the temperature to 45K (Buckyballs, 1991) . The only materials known to have higher transition temperatures than this new class of fullerene materials are ceramic "high temperature" superconductors.

Advantages of buckyball superconductors over existing ceramic materials are mainly in the relative ease with which the compounds can be formed into the shapes that are required, and the fact that they are isotropic, allowing current to flow equally well in any direction. The materials are metallic, and can be annealed, similar to older niobium superconductors. Niobium magnets must be cooled much closer to absolute zero than ceramic superconductors before they can be used. However, because of their better manufacturability and their considerably higher current carrying capabilities compared to ceramic materials, niobium magnets still make up almost all of the commercial market.
Another area of research is in organic superconductors, superconducting materials with zero metal content. This could be important for a number of reasons: lower toxicity, better raw materials availability, possibility of higher T_c's.

Results for fullerenes are very encouraging, but before useful applications are likely to emerge, the transition temperatures will have to rise, at least to the point where one may use liquid nitrogen instead of liquid helium to cool it (approximately 77 K). Theoretical analysis and some experimental data suggest that substituting larger buckyballs (higher order fullerenes) in place of C$_{60}$ will increase the transition temperature of the improved compounds (Argonne, 1991).

Possible applications include MRI magnets, mag-lev trains, high-speed computer chips based on the Josephson Junction and newer, even faster designs (Brown, 1991), long-distance power transmission, superconducting motors and generators, large magnets for physics research (e.g., superconducting supercollider), electronic shielding for supercomputers, and electronic devices based on SQUIDs (superconducting quantum interference devices).

10.7.6 High Strength Carbon Microfibers

Just months after the announced discovery of solid C$_{60}$, NEC

Corporation researcher Sumio IIjima revealed the existence of a new fullerene-related compound, the so-called "buckytube" (Iijima, 1991). These tubes are formed in tiled hexagonal arrays on the negative electrode of a DC carbon arc apparatus during the manufacture of raw soot. *Scientific American* describes the tubes as "thinner, more perfect in their molecular structure, more resistant to chemical attack and almost certainly stronger than any other fiber" (Ross, 1991). Recent work suggests that buckytubes have unusual electronic structures, and depending on their geometries, could behave as metals or semiconductors.

High performance fibers have revolutionized structural designs requiring very high strength-to-weight ratios, particularly in applications where one also needs an ability to withstand high temperatures, or to control electromagnetic properties of the material. Specialty fiber materials include those made from aramids (such as Nomex and Kevlar) and other polymers, boron, ceramic, metal, quartz, TFE-fluorocarbon, and graphite.
Graphite fibers already offer a tremendous range of strength, stiffness, and temperature resistance. Theory predicts that the "beam rigidity" of these carbon tubules exceeds the highest values found in presently available materials. Composite materials made from graphite fibers find wide application in military and aerospace markets, and more recently have infiltrated consumer product designs for tennis rackets, fishing poles, golf clubs, and automobile components.

Efficient production of buckytube "whiskers," very short carbon fibers, is already possible using the same apparatus which manufactures buckyballs. If the new discovery ultimately leads to a whole class of high performance carbon fibers which offer complementary or better characteristics than today's tremendously successful graphite fibers, the development could make possible even stronger and lighter structures. Development of manufacturing methods for specialty fibers typically require enormous investments in R&D before one can reliably control the quality and consistency of the fiber properties, however, and one may assume that this will be particularly true of buckytube fibers.

10.7.7 Nonlinear Optical Devices

At the 1991 Materials Research Society Fall Meeting in Boston,

Northwestern University researchers (also active in the CVD diamond precursor work described earlier) disclosed the potential usefulness of C_{60} films as nonlinear optical devices (Lasser Focus World, 1992) with high "second-order nonlinear optical susceptibility." An input light pulse from a Nd:YAG laser at 1064 nm doubled its frequency by passing through a thin film of C_{60} in an electric field. At an applied voltage of approximately 4 KV, the efficiency increased to nearly 10 times that of quartz. This property has significant and near-term potential applications in laser optical communications and optical computer designs.

A disclosure also was made by Hughes Research in the field of NLO: *solutions of C_{60} and C_{70} behave as "optical limiters".* The solutions allow low intensity light to pass through, but become progressively more opaque as the intensity increases. The saturation threshold they observed is equal to or better than any other known optical limiting material (Hughes Reports, 1992).

Many questions yet remain unanswered regarding fullerene-based nonlinear optical materials' abilities to compete effectively with other NLO materials already available. These include measurement of the magnitude and characterization of the nonlinear response of the materials. Recent findings show that many of the higher molecular weight fullerenes are chiral. Such chirality implies a strong likelihood of nonlinear optical response. Thus, one technical hurdle to be overcome to encourage further breakthroughs in higher order nonlinear effects, is in the perfection of high yield methods for the synthesis and separation of the large MW species.

10.7.8 Conclusions

The scope of applications proposed for buckminsterfullerenes and the family of derivative materials which they make possible, is staggering. The significance, from a scientific perspective, of research performed since Huffman and Krätschmer first disclosed their breakthrough process, is unquestionable. Although the jury is still out on the magnitude of the commercial potential for these discoveries, technological risk is very low that none of these diverse applications will prove to be economically important.

The direction which the fullerene industry will take from here depends on a number of issues including the resolution of patent applications pending, both on the composition of matter for fullerenes, and on the processes that have been developed to manufacture them. Exactly which manufacturing process will ultimately prove to be preferred for economical large-scale fullerene synthesis is not yet entirely clear. Unit volumes now being sold are still too low to justify investment in improved systems allowing reduction in the still-high labor content of fullerene products.

Today, Fullerene Science and technologies have demonstrated technical feasibility to displace existing products in scores of applications in wide-ranging markets totaling tens of billions of dollars. The extent to which fullerene products will be able to penetrate those markets varies meaningfully from one application to another, but as a group, the conclusion is inescapable: *buckyballs will keep on rolling*.

REFERENCES

Alberts A, Bray D, Lewis J, Raff M, Roberts K & Watson JD, <u>Molecular Biology of the Cell, Second Edition</u>, Garland Publishing, New York, 1989.

Albrecht-Buehler G, Is the cytoplasm intelligent too? *Cell and Muscle Motility* 6:1-21, 1985.

Allemand PM, Koch A, Wudl F, Rubin Y, Diederich F, Alvarez MM, Anz SJ & Whetten RL, Two different fullerenes have the same cyclic voltammetry. *J Am Chem Soc* 113:1050, 1991.

Allison AC & Nunn JF, Effects of general anaesthetics on microtubules. A possible mechanism of anaesthesia. *Lancet* 2:1326-1329, 1968.

Altman E & Colton R, Characterization of the interaction of C_{60} with Au(111). <u>Atomic and Nanoscale Manipulation of Materials</u>. Ed. Ph. Avoris, NATO ASI series, 1993.

Alvarez J & Ramirez BJ, Axonal microtubules: their regulation by the electrical activity of the nerve. *Neurosci Lett* 15:19-22, 1979.

Amit DJ, <u>Modeling Brain Function, The World of Attractor Neural Networks</u>, Cambridge University Press, Cambridge, MA, 1989.

Amos LA & Klug A, Arrangement of subunits in flagellar microtubules. *J Cell Sci* 14:523-550, 1974.

Amrein M, Stasiak A, Gross H, Stoll E, Travaglini G, Scanning tunneling microscopy for recA-DNA complexes coated on a conducting film. *Science* 240:514-516, 1988.

Aoki C & Siekevitz P, Plasticity in brain development. *Sci Am* (December 1988), 34-42.

Applewhite PB & Gardner FT, Tube escape behavior of paramecia. *Behav Biol* 9:245-250, 1973.

Argonne: A trend in T_c observed, *Diamond Depositions: Science and Technology*, August, 1991, p. 18 for a discussion of the correlation between the predicted lattice constant and the observed T_c.

Arnold SE, Lee VMY, Gur RE & Trojanowski JQ, Abnormal expression of two microtubule-associated proteins (MAP2 and MAP5) in specific subfields of the hippocampal formation in schizophrenia. *Proc Natl Acad Sci* 88:10850-10854, 1991.

Aszodi A, Muller V, Friedrich P & Spatz HC, Signal convergence on protein kinase A as a molecular correlate of learning. *Proc Natl Acad Sci* 88:5832-5836, 1991.

Atema J, Microtubule theory of sensory transduction. *J Theor Biol* 38:181-190, 1973.

Audenaert R, Heremans L, Heremans K & Engleborghs Y, Secondary structure analysis of tubulin and microtubules with Raman spectroscopy. *Biochim Biophys Acta* 996:110-115, 1989.

Baird B & Eeckman F, A neural network computer architecture for computation with oscillatory and chaotic attractors. *Proc 2nd Int Conf Fuzzy Logic and Neural Networks*, Iizuka, Japan, July 1992.

Balakrishnan AV, Introduction to Optimization Theory in a Hilbert Space, Springer-Verlag, New York, 1971.

Barto A, Sutton RJ & Anderson C, Neuron-like adaptive elements that can solve difficult learning control problems. *IEEE Trans. Systems, Man and Cybernetics*, Vol. SMC-13, pp. 834-846, 1983.

Becker RS, Golovchenko JA and Swartzentruber BS. Atomic-scale surface modifications using a tunnelling microscope. *Nature*, 325,29:419-421, 1987.

Becker JS, Oliver JM & Berlin RD, Fluorescence techniques for following interactions of microtubule subunits and membranes. *Nature* 254:152-154, 1975.

Bennett C, Brossand G & Ekert A, Quantum Cryptography. *Scientific American*, 267(4):50-57, 1992.

Bensimon G & Chernat R, Microtubule disruption and cognitive defects: effect of colchicine on learning behavior in rats. *Pharmacol Biochem Behavior* 38:141-145, 1991.

Bertsekas DP, Dynamic Programming: Deterministic and Stochastic Models, Prentice-Hall, New York, 1987.

Bethune DS, Meijer, Tang CW & Rosen JH, The vibrational Raman spectra of purified solid films of C_{60} and C_{70}. *Chem Phys Lett* 174, 3-4:219-222, 1990.

Bethune DS, Meijer G, Tang CW, Rosen JH, Golden WG, Seki H, Brown CA, and De Vries M, Vibrational Raman and infrared spectra of chromatographically separated C_{60} and C_{70} Fullerene clusters.*Chem Physic Letters* 179, 1-2:181-186, 1991.

Betzig E, Lewis A, Harootunian A, Isaacson N & Krätschmer E. Near-Field scanning optical microscopy (NSOM), development and biological applications. *Biophysics J* 49:269-279, 1986.

Bigot D & Hunt SP, Effect of excitatory amino acids on microtubule-associated proteins in cultured cortical and spinal neurons. *Neurosci Lett* 111:275-280, 1990.

Binnig G & Rohrer H, Scanning tunneling microscopy. *Helvetica Physica Acta* 55:726-735, 1982.

Binnig G and Rohrer H, Scanning tunneling microscopy. *Surface Science* 152/153:17-26, 1985.

Bloembergen N & Zewail AH, Energy redistribution in isolated molecules and the question of mode-selective laser chemistry revisited. *J Physical Chemistry* 88:5459-65, 1984.

Boulas P, Kutner W & Kadish KM, Bucktube(basket)ball: water solubilization of C_{60} by means of cyclodextrin inclusion chemistry, presented at <u>Fullerenes: chemistry, physics and new directions symposium of the 181st meeting of the electrochemical society</u>, St. Louis, Missouri, May 17-22, 1992.

Bornens M. The centriole as a gyroscopic oscillator. Implications for cell organization and some other consequences. *Biol Cellulaire* 35:115-132, 1979.

Brown, Chapell. Despite problems, superconducting logic thrives. *Electronic Engineering Times*, Sept. 23, 1991, p. 43.

Bryant A, Smith DPE, Binnig WA, Harrison WA and Quate CF. Anomalous distance dependence in scanning tunneling microscopy. *Appl Phys Lett* 49(15):936-938, 1986.

Buckyballs: Wide open playing field for chemists, *Science* 254:1706, 1991.

Buseck PR, Tsipursky and Hettich R. Fullerenes from the geological environment. *Science* 257:215-217, 1992.

Butkovskii AG, <u>Distributed Parameter Systems</u>, Elsevier, New York, 1967.

Butkovskii AG & Samoilenko Y, *Automation and Remote Control* A485, 1979.

Carter FL, The molecular device computer:point of departure for large scale cellular automata. *Physica* 10D:175-194, 1984.

Chabre Y, Djurado D, Armand M, Romanow WR, Constel N, McCanley JP, Fischer JE & Smith AB, Electrochemical intercalation of lithium into solid C_{60}. *J Am Chem Soc* 114:764, 1992.

Chabre Y, Djurado D, Barral M, Electrochemical study of the sodium/solid C_{60} system, Material Research Society extended abstract A Boston Meeting, Nov. 30-Dec 4, 1992.

Churchland PS & Sejnowski TJ, Perspectives on cognitive neuroscience. *Science* 242:741-745, 1988.

Cohen J, Beisson J, The cytoskeleton. In <u>Paramecium</u>, ed. HD Gortz, Springer-Verlag, Berlin, 363-392, 1988.

Compton RN, Hettich RL, Britt P, Puretzky AA, Frey WF, Tuinman AA, Adcock JL and Mukherjee P, On the generation, separation, physics and chemistry of carbon clusters. *J Electrochem Soc* 139:239c, 1992.

Connolly S, Nishimura D and Macovski A. *IEEE Trans. on Medical Imaging* 5:106, 1986.

Conrad M, Molecular automata. In: <u>Lecture Notes in Biomathematics, Vol.4: Physics and Mathematics of the Nervous System</u>, eds. M Conrad, W Guttinger and M Dal Cin, Springer-Verlag, Heidelberg, 419-430, 1974.

Conrad M, Microscopic - macroscopic interfaces in biological information processing. *Biosystems*, 16:345, 1984.

Conrad M, On design principles for a molecular computer. *Commun ACM* 28:464-480, 1985.

Cronly-Dillon J, Carden D & Birks C, The possible involvement of brain microtubules in memory fixation. *J Exp Biol* 61:443-454, 1974.

Crowther RA, Finch JT and Pearse BMF, On the structure of coated vesicles. *J Mol Biol* 103:785-798, 1976.

Curie MP, Sur la symétrie dans les phenomenes physiques, symetrie d'un champ électrique et d'un champ magnetique. *J de Physique* Ser 3: 393-417, 1894.

Curl RF & Smalley RE, Fullerenes. *Scientific American*, 54-63, October 1991.

Dahleh M, Peirce AP & Rabitz H, Optimal control of uncertain quantum systems. *Phys Rev A* 42(3):1065, 1990.

Davies EV, *IEEE Trans. on Information Theory* I2-23:530-34, 1977.

De Callatay AM, <u>Natural and artificial intelligence. Processor systems compared to the human brain</u>. North-Holland, Amsterdam, 1986.

De Freitas L. Notes on some pentagonal "mysteries" in Egyptian and Christian iconography in <u>Fivefold Symmetry</u> ed I. Hargittai World Scientific Singapore, 1992 pp. 307-332.

Delfour M & Mitter SK, *SIAM J on Control* 10:329, 1972.

Desmond NL & Levy WB, Anatomy of associative long-term synaptic modification. In: <u>Long-Term Potentiation: From Biophysics to Behavior</u>, eds. PW Landfield and SA Deadwyler, 1988, pp.265-305

Deutsch D, Quantum computational networks. *Proc Royal Society A* 425:73-82, 1989.

Deutsch D, Quantum computation, *Physics World,* June, 57-61, 1992.

Devoret MH, Esteve D, Urbina C, Single-electron transfer in metallic nanostructures, *Nature* 360:547-553, 1992.

Dewdney AK, Computer recreations. *Sci Am* 252:18-30, 1985.

Diederich F & Whetten RL, Beyond C_{60}: the higher Fullerenes. *Acc Chem Res* 25:119-126, 1992.

Dietrich HP, Lanz M & Moore DF, Ion beam machining of very sharp points. *IBM Tech Disclosure Bul*, 27(5):3039-3040, 1984.

Dorf R, <u>Modern Control Systems</u>, Addison-Wesley, Reading, MA,1992.

Drake BR, Sonnenfeld J, Schneir PK, Hansma G, Solugh RV & Coleman. A tunneling microscope for operation in air or fluids. *Rev Sci Instr* 57(3):441-445.

Dresselhaus MS, Down the straight and narrow. *Nature* 358:195-196, 1992.

Drexler KE, Molecular Engineering: an approach to the development of general capabilities for molecular manipulation. *Proc Natl Acad Sci USA* 78(9): 5275-5278.

Drexler KE, Engines of Creation. Garden City, Anchor Press, 1986.

Dryl S, Behavior and motor responses in Paramecium. In: Paramecium--A Current Survey, ed. WJ Van Wagtendonk, Amsterdam, Elsevier, pp.165-218, 1974.

Dubois D, Kadish KM, Flanagan S, Haufler RE, Chibante LPE & Wilson LJ, Spectro-electrochemical study of the C_{60} and C_{70} fullerenes and their mono-, di-, tri-, and tetra-ions. *J Am Chem Soc* 113:4364, 1991a.

Dubois D, Kadish KM, Flanagan S, Wilson J, Electrochemical detection of fulleronium and highly reduced fullerides (C_{60}^{5-}) ions in solution. *J Am Chem Soc* 113:7773, 1991b.

Dustin P, Microtubules 2nd Revised Ed, Springer, Berlin, 1984.

Ebbeson TW & Ajayan PM, Large-scale synthesis of carbon nanotubes. *Nature* 358:220-22, 1992.

Eccles JC, Evolution of consciousness. *Proc Natl Acad Sci* 89:7320-7324, 1992.

Eccles JC, Do mental events cause neural events analogously to the probability fields of quantum mechanics? Proc R Soc London Ser B 277:411-428, 1986.

Elser V, Haddon RC, Icosahedral C_{60}: an aromatic molecule with a vanishingly small ring current magnetic susceptibility. *Nature* 325:792-794, 1992.

Elser V, & Haddon RC, Magnetic behavior of icosahedral C_{60}. *Physical Review* A 36(10):4579-4584, 1987.

Engelborghs Y, Dynamic aspects of the conformational states of tubulin and microtubules. *Nanobiology* 1:97-105, 1992.

Erickson RO, Tubular packing of spheres in biological fine structure. *Science* 181:705-781, 1973.

Ettinger RCW, Man into Superman. New York, St. Martin's, 1972.

Fasman GD, Prediction of protein structures and the principles of protein conformation. Plenum Press, New York, 1989.

Feynman RP, There's plenty of room at the bottom (Chapter 16 in Gilbert, 1961).

Findeisen W, Bailey FN, Bradys M, Malinowski K, Tatjewski P & Wozniak A, Control and Coordination in Hierarchical Systems, Wiley, 1980.

Fisher CJ & Fuller H, How to inscribe a dodecahedron in a sphere, in Fivefold Symmetry ed I Hargittai, World Scientific, Singapore, 1992, pp. 167-170.

Friedrich P, Protein structure: the primary substrate for memory. *Neurosci* 35:1-7, 1990.

Friedrich P, Fesus L, Taresa E & Czeh G, Protein cross-linking by transglutamination induced in long-term potentiation in the CA1 region of hippocampal slices. *Neurosci* 43:331-334, 1991.

Fröhlich H, Long range coherence and the actions of enzymes. *Nature* 228:1093, 1970.

Fröhlich H, The extraordinary dielectric properties of biological materials and the action of enzymes. *Proc Natl Acad Sci* 72:4211-4215, 1975.

Fröhlich H, Coherent excitations in active biological systems. In: Modern Bioelectrochemistry, eds. F Guttmann and H Keyzer, Plenum Press, New York, 241-261, 1986.

Frum CI, Engleman R Jr, Hedderich HG, Bernath PF, Lamb LD & Huffman DR, The infrared emission spectrum of gas-phase C_{60}. *Chem Phys Lett* 176:506-508, 1991.

Fukui K & Asai H, Spiral motion of Paramecium caudatum in a small capillary glass tube. *J Protozool* 23:559-563, 1976.

Garroway A, Grannell P & Mansfield P. *J of Physics* C7:L457, 1974.

Geerts H, Nuydens R, Nuyens R, Cornelissen F, De Brabander M, Pauwels P, Janssen PAJ, Song YH & Mandelkow EM, Sabeluzole, a memory-enhancing molecule, increases fast axonal transport in neuronal cell cultures. *Experimental Neurology* 117:36-43, 1992.

Gelber B, Retention in Paramecium aurelia. *J Comp Physiol Psych* 51:110-115, 1958.

Gelfand IM & Fomin SV, Calculus of Variations, Prentice-Hall, Englewood Cliffs, NJ, 1963.

Gilbert HD, Miniaturization Reinhold, New York, 1961.

Gutmann F, Some aspects of charge transfer in biological systems. In: Modern Bioelectrochemistry, eds. F Gutmann & H Keyzer, Plenum Press, New York, 177-197, 1986.

Haddon RC, π-electrons in three dimensions Acc Chem Res 21:243-249, 1988.

Haddon RC, Brus LE & Raghavachari K, Electronic structure and bonding in icosahedral C_{60}. Chemical Phys Letters 125:5-6, 459-464, 1986.

Halpain S & Greengard P, Activation of NMDA receptors induces rapid dephosphorylation of the cytoskeletal protein MAP2. Neuron 5:237-246, 1990.

Hameroff SR, Ultimate Computing: Biomolecular Consciousness and Nanotechnology, North-Holland, Amsterdam, 1987.

Hameroff S, Dayhoff JE, Lahoz-Beltra R, Rasmussen S, Insinna E & Koruga D, Nanoneurology and the cytoskeleton: quantum signaling and protein conformational dynamics as cognitive substrate. In: Behavioral Quantum Neurodynamics, ed. K Pribram & H Szu, Pergamon Press, in press.

Hameroff SR, Dayhoff JE, Lahoz-Beltra R, Samsonovich A & Rasmussen S, Models for molecular computation: conformational automata in the cytoskeleton. IEEE Computer (Special Issue on Molecular Computing), 30-39, 1992.

Hameroff SR, Rasmussen S & Mansson B, Molecular automata in microtubules: basic computational logic for the living state? In: Artificial Life, the Santa Fe Institute Studies in the Sciences of Complexity, Vol. VI, ed. C Langton, Addison-Wesley, Reading, MA, 521-553, 1989.

Hameroff SR, Smith SA & Watt RC, Automaton model of dynamic organization in microtubules. Ann NY Acad Sci 466:949-952, 1986.

Hameroff SR, Smith SA & Watt RC, Nonlinear electrodynamics in cytoskeletal protein lattices. In: Nonlinear Electrodynamics in Biological Systems, eds. WR Adey and AF Lawrence, Plenum Press, New York, 567-583, 1984.

Hameroff SR & Watt RC, Do anesthetics act by altering electron mobility? *Anesth Analg* 62:936-940.

Hameroff SR & Watt RC, Information processing in microtubules. *J Theor Biol* 98:549-561, 1982.

Hameroff SR et al. Scanning tunneling microscopy of cytoskeletal proteins. *J Vacuum Sci & Tech* Al:607-611, 1990.

Hammermesh M. Group Theory and its Application to Physical Problems. Addison Wesley, Reading, MA 1972.

Hansma PK and Tersoff J. Scanning Tunneling Microscopy. *J Appl Phys* 61(2):R1-R23.

Hare JP, Kroto HW & Taylor R, Preparation and UV/visible spectra of Fullerenes C_{60} and C_{70}. *Chem Phys Lett* 177:394, 1991.

Hargittai I, Fivefold Symmetry. World Scientific, Singapore, 1992.

Harter WG & Reimer T, Rotation-vibration spectra of icosahedral molecules. III. Rotation energy level spectra for half-integral angular momentum icosahedral molecules. *J Chem Phys* 94(8):5426-5434, 1991.

Harter WG & Weeks DE, Rotation-vibration spectra of icosahedral molecules. I. Icosahedral symmetry analysis and fine structure. *J Chem Phys* 90(9):4724-4743, 1989.

Haufler RE, Conccicao J, Chibante LPF, Chai Y, Byrne NE, Flanagan S, Haley MM, O'Brian SC, Pan C, Ziao Z, Billups WE, Cinufolini MA, Hauge RH, Margrave JL, Wilson LJ, Curl RF & Smalley RE, Efficient production of C_{60} (Buckminsterfullerene), $C_{60}H_{36}$ and the solvated buckide ion. *J Phys Chem* 94:8634, 1990.

Haufler RE, Chaie Y, Chibante LPF, Conceicao J, Jin C, Wang S, Maruyama S & Smalley RE. *Materials Research Society Symposium Proceedings*, Eds. RS Averback, J Bernholc & DL Nelson, Materials Research Society, Boston, MA, 627, 1991.

Hayden GW & Mele EJ, π bonding in the icosahedral C_{60} cluster. *Phys Rev B* 36,9:5010-5015, 1987.

Hebard AF, Superconductivity in doped fullerenes. *Physics Today* 11:26-32, 1992.

Hebb DO, The Organization of Behavior, Wiley, New York, 1949.

Hecht-Nielsen R, Neurocomputing, Addison-Wesley, 1990.

Heuser J & Kirchhausen T, Deep-etch views of clathrin assemblies. *J Ultrastructure Res* 92:1-27, 1985.

Hirokawa N, Molecular architecture and dynamics of the neuronal cytoskeleton. In: The Neuronal Cytoskeleton, ed. RD Burgoyne, Wiley Liss, 5-74, 1991.

Hirsch MW, Convergent activation dynamics in continuous-time networks. *Neural Networks* 2:331-349, 1989.

Ho YC & Mitter SK (Eds), Directions in Decentralized Control, Many-Person Optimization and Large-Scale Systems, Plenum Press, 1976.

Holme TA & Hutchinson JS, A theory for long-lived high-energy excitation of a local mode using two lasers. *Chem Phys Letters* 124:181-186, 1986.

Holme TA & Hutchinson JS, A theoretical application of coherent multicolor laser spectroscopy to selective control of singlet and triplet excitations in carbon monosulfide. *J Chem Phys* 86:42, 1987.

Hopfield JJ, Neural networks and physical systems with emergent collective computational abilities. *Proc Natl Acad Sci* 79:2554-2558, 1982.

Hotani H, Lahoz-Beltra R, Combs B, Hameroff S & Rasmussen S, Microtubule dynamics, liposomes and artificial cells: in vitro observation and cellular automata simulation of microtubule assembly/disassembly and membrane morphogenesis. *Nanobiology* 1:61-74, 1992.

Hoult D, The solution of the Bloch equation in the presence of a varying B1 field- an approach to selective pulse analysis . *J of Magnetic Resonance* 35:69-86, 1979.

Howard JB, McKinnon JT, Marovsky Y, Lafleur A & Johnson ME. *Nature* 352:139-141, 1991.

Huang GM, Tarn TJ & Clark JW. *J Math Phys* 24:2608, 1983.

Huffman DR, Solid C_{60}. *Physics Today*, 44:22-29, 1991.

Hughes Reports C_{60} and C_{70} are optical limiters. *Diamond Depositions: Science and Technology*. March 30, 1992, p. 12.

Iijima S, Direct observation of the tetrahedral bonding in graphitized carbon black by high resolution electron microscopy. *J Crystal Growth* 50:675-683, 1980.

Iijima S. Helical microtubules of graphitic carbon. *Nature* 354:56, 1991.

Iijima S, Symposium on the Physics and Chemistry of Finite Systems, Richmond, Virginia, 8-12 October 1991.

Iijima S, Ichihashi T & Ando Y, Pentagons, heptagons and negative curvature in graphite microtubule growth. *Nature* 356: 7776-780, 1992.

Insinna E, Synchronicity and coherent excitations in microtubules. *Nanobiology* 1(2):191-208, 1992.

Jahn HA, Teller. Stability of polyatomic molecules in degenerate electronic states. *Proc Roy Soc* (London) 46:221-235, 1937.

Jakubetz W, Manz J & Schreier H, Theory of optimal laser pulses for selective transitions between molecular eigenstates. *Chem Phys Lett* 65(1), 1990.

Jehoulet C, Bard AJ & Wudle F, Electrochemical reduction and oxidation for C_{60} films. *J Am Chem Soc* 113:5456, 1991.

Jehoulet C, Obeng YS, Kim YT, Zhou F & Bard AJ, Electrochemistry and Langmuir trough studies of C_{60} and C_{70} films. *J Am Chem Soc* 114:4237, 1992.

Jiang Q, Xia H, Zhang Z & Tian D, Vibrational spectrum of C_{60}. *Chem Phys Lett* 192(1-3), 93-96, 1992.

Judd, BR, A crystal field of icosahedral symmetry. *Proc Roy Soc* (London) A241, 122-131, 1957.

Kamaseki T & Kadota K, The "vesicle in a basket". *J Cell Biol* 42:202-220, 1969.

Karakasoglu A, Sundareshan SI & Sundareshan MK, Neural network-based identification and control of nonlinear systems: A novel dynamical network architecture and training policy, Proceedings of the 30th IEEE Conference on Decision and Control, Brighton, England, 1991.

Kaufman RD, Biophysical mechanisms of anesthetic action: Historical perspective and review of current concepts. *Anesthesiology* 46:49-62,1977.

Keen J, Clathrin assembly proteins: affinity purification and a model for coat assembly. *J Cell Biol* 105:1989-1998, 1987.

Kelly WG, Passaniti A, Woods JW, Daiss JL, Roth TF, Tubulin as a molecular component of coated vesicles. *J Cell Bio* 97:1191-1199, 1983.

Klug DD, Howard JA & Wilkinson DA, The pressure dependence of the infrared active vibrations in C_{60}. *Chem Phys Lett* 188,3-4:168-170, 1992.

Koch A, Khemani KC & Wudl F, Preparation of Fullerenes with a simple benchtop reactor. *J Org Chem* 56:4543-4547, 1991.

Koruga D, Microtubule screw symmetry: packing of spheres as a latent bioinformation code. *Ann NY Acad Sci* 466:953-955, 1984.

Koruga D, Qi Engineering. Poslovna Politika, Belgrade, 1984.

Koruga D & Hameroff S, Neurocomputing and infinite-valued logic, Proceedings of the 2nd International Conference on Fuzzy Logic and Neural Networks (Iizuka, Japan), 177-180, 1992.

Koruga D, Hameroff S, Simic-Krstić, Janković S, Trifunović M. and Sundareshan M. Self-assembled molecular computer based on Fullerene C_{60}: From nanobiology to nanotechnology, Proceedings of the 4th Int. Sym. on Bioelectronic and Molecular Electronic Devices, R & D Assocation for Future Electronic Devices, Nov. 30-Dec. 2, 1992, pp. 1-5, Miyazaki, Japan.

Koruga D, Neurocomputing and consciousness. *Int J on Neural and Mass-Parallel Computing and Information Systems*, 32-38, 1991.

Koruga D, Neuromolecular computing. *Nanobiology* 1:5-24, 1992.

Koruga D, Simić-Kristić J, Semiconductor and crystal symmetry assessment of microtubule proteins as molecular machines. *J Mol Elec* 6:167-173, 1990.

Koruga D, Simić-Kristić J, Trifunović M, Janković S, Hameroff S, Withers J & Loutfy R, Imaging fullerene C_{60} with atomic resolution using a scanning tunneling microscope. *Fullerene Sciences and Technology* 1(1):93-100, 1993.

Kosko B, Neural Networks and Fuzzy Systems - A Dynamical Systems Approach to Machine Intelligence, Prentice-Hall, Englewood Cliffs, New Jersey, 1992.

Kosloff R, Rice S, Gaspard P, Tersigni S & Tannor D, Wavepacket dancing: achieving chemical selectivity by shaping light pulses. *Chem Phys* 139(1), 1989.

Krätschmer W, How we came to produce C_{60}-fullerite. *Z Phys D- Atoms, Molecules and Clusters* 19:405-408, 1991.

Krätschmer W, Lamb LD, Fostiropoulos K & Huffman DR, Solid C_{60}: A New Form of Carbon. *Nature* 347:354-358, 1990.

Krätschmer W, Fostiropoulos K & Huffman DR, The infrared and ultraviolet absorption spectra of laboratory-produced carbon dust: evidence for the presence of the C_{60} molecule. *Chem Phys Lett* 170:167-174, 1990.

Kroto HW, C_{60}: Buckminsterfullerene, the celestial sphere that fell to earth. *Angewandte Chemie* (Inf. edition in English), 31,2:111-246, 1992.

Kroto HW, Space, stars, C_{60} and soot. *Science* 242:1139-1145, 1988.

Kroto HW, Allaf AW, Balm SP, C_{60}: Buckminsterfullerene. *Chem Rev* 91:1213-1235, 1991.

Kroto HW, Heath JR, O'Brien SCO, Curl RF & Smalley RE, C_{60}: Buckminsterfullerene. *Nature* 318:162-163, 1985.

Kwak S & Matus A, Denervation induces long-lasting changes in the distribution of microtubule proteins in hippocampal neurons. *J Neurocytol* 17:189-195, 1988.

Lahoz-Beltra R, Hameroff S, & Dayhoff JE, Cytoskeletal logic: a model for molecular computation via Boolean operations in microtubules and microtubule-associated proteins. *BioSystems*, in press.

Lamb LD, Huffman DR, Workman RK, Howells S, Chen T, Sarid D & Ziolo RF. *Science*, 1992.

Landau ID, Adaptive Control-The Model Reference Approach, Marcel Dekker, New York, 1981.

Lasek RJ, The dynamic ordering of neuronal cytoskeletons. *Neurosci Res Prog Bull* 19:7-31, 1981.

Laser Focus World. Fullerene thin film shows high nonlinear optical susceptibility. January 1992, p.11.

Leach J & Sloane NJA, Sphere Packing and error-correcting codes. *Can J Math* 23, 4:718-745, 1971.

Lewis A, Isaacson M, Harootunian A & Muray A, Development of a 500 Å spatial resolution light microscope. *Ultramicroscopy* 13, 227-232, 1984.

Luenberger DG, Optimization by Vector-Space Methods, Wiley, New York, 1969.

Lukasiewiz J, Investigations in the sententi calculus (1930): in selected works, Amsterdam: Reidel, 1970.

Lund EE, A correlation of the silverline and neuromotor systems of Paramecium. *University of California Publication in Zoology* 39(2):35-76, 1933.

Lynch G & Baudry M, Brain spectrin, calpain and long-term changes in synaptic efficacy. *Brain Res Bull* 18:809-815, 1987.

Mandelkov E, Domains, antigenicity and substance of tubulin, in Microtubules and Microtubule Inhibitors (De Brabander M & De Mey JD eds) Elsevier, Amsterdam, 31-47, 1985.

Mandelkov E & Mandelkov EM, Tubulin, microtubules and oligomers: molecular structure and implications for assembly. *Cell Movement* 2:23-45, 1989.

Manka R & Ogrodnik B, Soliton model of transport along microtubules. *J Biol Physics*, in press.

Maren AJ, Handbook of Neural Computing Applications, Academic Press, 1990.

Margulis L, Origin of Eukaryotic Cells. New Haven, Yale University Press, 1975.

Margulis L & Sagan D, Strange fruit on the tree of life. *The Sciences*, May/June, 38-45, 1986.

Margulis L, To L & Chase D, Microtubules in prokaryotes. *Science* 200:1118-1124, 1978.

Marks R & Fuller BR. The Dymaxion World of Buckminster Fuller. Anchor Books, New York, 1973.

Mascarenhas S, The electret effect in bone and biopolymers and the bound water problem. *Ann NY Acad Sci* 238:36-52, 1974.

Matsumoto G & Sakai H, Microtubules inside the plasma membrane of squid giant axons and their possible physiological function. *J Membr Biol* 50:1-14, 1974.

Matsuyama SS & Jarvik LF, Hypothesis: microtubules, a key to Alzheimer's disease. *Proc Natl Acad Sci* 86:8152-8156, 1989.

McLellan AG, Eigenfunctions for integer and half-odd integer values of J symmetrized according to the icosahedral group and the group C_{32}. *J Chem Physics* 34(4):1350-1359, 1961.

Milch JR, Computer based on molecular implementation of cellular automata in: Molecular Electronic Devices: Proceedings of the Third International Symposium on Molecular Electronic Devices, Oct 6-8, 1986 Arlington, VA. FL Cartes & H Wohltjen eds, 1987 Elsevier. North-Holland 1988.

Mileusnic R, Rose SP & Tillson P, Passive avoidance learning results in region specific changes in concentration of, and incorporation into, colchicine binding proteins in the chick forebrain. *Neur Chem* 34:1007-1015, 1980.

Minsky M, The Society of Mind. Simon and Schuster, New York, 1986.

Miranda RN, Garcia A, Baro R, Garcia R, Pena JL & Rohrer H, Technological applications of scanning tunneling microscopy at atmospheric pressure. *Appl Phys Lett* 47(4), 1985.

Mithieux G, Roux Bernard & Rousset B, Tubulin-chromatin interactions: evidence for tubulin binding sites on chromatin and isolated oligonucleosomes. *Biochemica et Biophysica Acta* 888:49-61, 1986.

Montgomery EB, Clare MH, Sahrmann S, Buchholz SR, Hibbard LS & Landau WM, Neuronal multipotentiality: evidence for network representation of physiological function. *Brain Research* 580:49-61, 1992.

Moravec H, Mind Children. San Francisco, University of California Press, 1987.

Morton JR, Preston KF, Krusic PJ, Hill SA & Wasserman E. The dimerization of RC_{60} radicals. *J Am Chem Soc* 114:5454-5455, 1992.

Moshkov DA, Savaljeva LN, Yanjushina GV, Funtikov VA, Structural and chemical changes in the cytoskeleton of the goldfish Mauthner cells after vestibular stimulation. *Acta Histochemica Suppl-Band* XLI, S:241-247, 1992.

Mushin I, Sundareshan MK, Sudharsanan SI & Karakasoglu A, Adaptive excitation and governor control of synchronous generators using multilayer recurrent neural networks. Proceedings of the 31st IEEE Conference on Decision and Control, Tucson, AZ, 1992.

Narendra KS & Annaswamy A, Stable Adaptive Systems, Prentice-Hall, Englewood Cliffs, NJ, 1989.

Negri F & Orlandi G, Quantum-chemical investigation of Franck-Condon and Jahn-Teller activity in the electronic spectra of Buckminsterfullerene. *Chem Phys Lett* 144:31-37, 1988.

Negri F, Orlandi G & Zerbetto F, The infrared and Raman active vibrational frequencies of C_{60} hexaanion, *Chem Phys Lett* 196:3-4, 1992.

Nicolas G & Prigogine I, Self Organization in Nonequilibrium Systems. New York, Wiley, 1977.

Njegoš PP, The Ray of the Microcosm. Vajat, Belgrade, 1989.

Nye JF, Physical Properties of Crystals. Clarendon Press, Oxford, 1957.

Ochs S, Axoplasmic Transport and Its Relation to Other Nerve Functions, Wiley-Interscience, New York, 1982.

O'Neill JJ. Prodigal Genius: The Life of Nikola Tesla, Ives Washburn, New York, 1945.

Packard N, Adaption towards the edge of chaos. Preprint, 1988.

Pang LSK, Vassallo AN & Wilson MN, Fullerenes from coal. *Nature* 352:480, 1991.

Parducz B, On a new concept of cortical organization in Paramecium. *Acta Biol Acad Sci Hung* 13:299-322, 1962.

Park SM, Lu S-P & Gordon RJ, Coherent laser control of the resonance-enhanced multiphoton ionization of HCl, *J Chem Phys* 94(2;2), 1991.

Parker DH, Wurz P, Chatterjee K, Lykke KR, Hunt JE, Pellin MJ, Hemminger JC, Gruen DM & Stock LM, High yield synthesis and mass-spectrometric characterization of Fullerene C_{60} to C_{266}. *J Am Chem Soc* 113:7499-7503, 1991.

Parker DH, et al, Fullerenes and giant Fullerenes: synthesis, separation and mass spectrometive characterization, submitted to *Carbon* September 4, 1992, for publication in a special issue devoted to Fullerenes.

Pasquarello A, Schlüter M & Haddon RC, Ring currents in icosahedral C_{60}. *Science* 257:1660-1661, 1992.

Pearce B, Clathrin: A unique protein associated with intracellular transfer of membrane coated vesicles. *Proc Natl Acad Sci, USA* 73(4) 1255-1259, 1976.

Penrose R, The Emperor's New Mind: Concerning Computers Minds and the Laws of Physics, Oxford University Press, 1989.

Peirce AP, Dahleh M & Rabitz H, Optimal control of quantum-mechanical systems: existence, numerical approximation, and applications, *Phys Rev A* 37(12) 4950-4964, 1988.

Perelson AS, Immune network theory, *Immunological Reviews* 110:5-35, 1989.

Peters G & Jansen M, *Agew Chem Int Ed Eng* 31:223, 1992.

Pohl DW, Denk W & Lanz M, Optical stethoscopy: Image recording with resolution. *Appl Phys Lett* 44(7) 651-653, 1984.

Popper KR & Eccles VC, The Self and Its Brain, Springer, Berlin, 1977.

Pradeep T & Rao CNR, *Materials Research Bulletin* 26:1101, 1991.

Pribram KH, Brain and Perception: Holonomy and Structure in Figural Processing. Hillsdale, NJ, Lawrence Erlbaum, 1991.

Puiseux-Dao S, Environmental signals and rhythms on the order of hours - role of cellular membranes and compartments and the cytoskeleton. In: Cell Cycle Clocks, ed. LN Edmunds Jr, Marcel-Dekker, New York, 351-363, 1984.

Rabitz H, Systems analysis at the molecular scale. *Science* 246:221-226, 1989.

Rakočević M. Three-dimensional model of the genetic code. *Acta Biol Eap* 13:109-116, 1988.

Rasenick MM, Wang N & Yan K, Specific associations between tubulin and G proteins: participation of cytoskeletal elements in cellular signal transduction. In:

The Biology and Medicine of Signal Transduction, ed. Y Nichizuka et al, Raven Press, New York, 381-386, 1990.

Rasmussen S, Karampurwala H, Vaidyanath R, Jensen KS & Hameroff S, Computational connectionism within neurons: a model of cytoskeletal automata subserving neural networks. *Physica D* 42:428-449, 1990.

Raynal J., Determination of point group harmonics for arbitrary J by a projection method. II. Icosahedral group, quantization along an axis of order 5. *J Math Phys* 25(5) 1187-1194, 1984.

Reeke GR & Edelman GM, Selective networks and recognition automata.   In: Computer Culture: The Scientific, Intellectual, and Social Impact of the Computer, ed. HR Pagels, Ann NY Acad Sci 426:181-201, 1984.

Reguiero MN, Monceau P & Hodeau JL, Crushing C_{60} to diamond at room temperature. *Nature* 16:237-239, 1992.

Ringger M, Hidber HR, Schlögl, Oelhafen P & Güntherodt. Nanometer lithography with the scanning tunneling microscope. *Appl Phys Lett* 46(9) 832-834, 1985.

Roberts MR, Serendipity: Accidental Discoveries in Science. John Wiley & Sons, New York, 1989.

Ross PE. Buckytubes. *Scientific American* p. 24, Dec. 1991.

Sabry JH, O'Connor TP, Evans L, Toroian-Raymond A, Kirschner M & Bentley D, Microtubule behavior during guidance of pioneer neuron growth cones in situ. *J Cell Biol* 115(2) 381-395, 1991.

Saito S & Oshiyama A, Cohesive mechanism and energy bands of solid C_{60}. *Physical Review Letters* 66,20, 2637-2640, 1991.

Samsonovich A, Scott A & Hameroff SR, Acousto-conformational transitions in cytoskeletal microtubules: implications for neuro-like protein array devices. *Nanobiology* 1(4) 457-468, 1992.

Sataric MV, Zakula RB & Tuszynski JA, A model of the energy transfer mechanisms in microtubules involving a single soliton, *Nanobiology* 1(4)445-456, 1992.

Satpathy S. Electronic structure of the truncated icosahedral C_{60} cluster. *Chemical Physical Letters* 130(6), 1986.

Saunders M, Buckminsterfullerene: the inside story. *Science* 213:330-331, 1991.

Schectman IA, Blech D, Gratias D & Cahn JW, Metallic phase with long-range orientational order and no translational symmetry. *Phys Rev Lett* 53(20):1951-1953,1984.

Schmid S, The mechanism of receptor-medicated endocytosis: more questions than answers. *Bio Essays* 14(9) 589-596, 1992.

Schrodinger E. *What is Life and Mind and Matter?* Cambridge University Press, Cambridge, 1967.

Schulman H & Lou LL, Multifunctional Ca^{2+}/calmodulin-dependent protein kinase: domain structure and regulation. *Trends Biochem Sci* 14:62-66, 1989.

Schneiker CW and Hameroff SR, Nanotechnology workstations based on scanning tunneling/optical microscopy: Applications to molecular scale devices. In: Molecular Electronic Devices: Proceedings of the Third International Symposium on Molecular Electronic Devices, Oct. 6-8, 1986, Arlington, VA. F L. Carter and H. Wohltjen, eds. (1986), Elsevier North-Holland, pp. 69-90, 1988.

Scott AC, Neurophysics, Wiley, New York, 1977.

Scott AC, Solitons and bioenergetics. Nonlinear Electrodynamics in Biological Systems, Eds. WR Adey & AF Lawrence, New York, Plenum Press, 1984.

Segar L, Wen LQ & Schienff JB, Prospects for using C_{60} and C_{70} in lithium batteries. *J Electrochem Soc* 128, 1991.

Shi S, Woody A & Rabitz H, Optimal control of selective vibrational excitation in harmonic linear chain molecules, *J Chem Phys* 88(11) 6870, 1988.

Shinohara H, Sato Y, Takayama M, Izuoka A & Sugawara T, *J Phys Chem* 95:8449, 1991.

Shapiro M & Brumer P, Laser control of product quantum state populations in unimolecular reactions, *J Chem Phys* 84(7) 4103-4104, 1986.

Shaskolskaya S, Fundamentals of Crystal Physics. Mir Publishers, Moscow, 1982.

Shoulders KR. Toward complex systems. In: Microelectronics and Large Systems. Spartan Books, 97-128, 1965.

Silva AV, Stevens CF, Tonegawa S & Wang, Deficient hippocampal long-term potentiation in α-calcium-calmodulin kinase II mutant mice. *Science* 257:201-206, 1992.

Silva AV, Paylor R, Wehner JM & Tonegawa S, Impaired spatial learning in α calcium calmodulin kinase II mutant mice. *Science* 257:206-211, 1992.

Simić-Krstić J, Kelley M, Schneiker L, Krasovich M, McCuskey R, Koruga D, Hameroff S, Direct observation of microtubules with the scanning tunneling microscope. *FASEB J* 3:2186-2192, 1989.

Simić-Krstić J. Raman spectroscopy of microtubule proteins. Proceedings of the 2nd International Conference on MEBC, Moscow, 1989, Kluwer, Dordrecht, 123-124, 1991B.

Sinha K, Menendez, Hanson RC, Adams GB, Page JB, Sankey OF, Lamb LD & Huffman DR, Evidence for solid-state effects in the electronic structure of C_{60} films: a resonance- Raman study, *Chem Phys Lett* 186, 2-3:287-290, 1991.

Sirontin I & Shaskolsky MP. Fundamentals of Crystal Physics. Nauka, Moscow (in Russian), 1979.

Sloane NJA, The packing of spheres. *Scientific American* 1:116-125, 1984.

Slotine J & Li WP, Applied Nonlinear Control, Prentice-Hall, Englewood Cliffs, NJ, 1991.

Smalley RE, Great balls of carbon *The Sciences* 3/4:22-28, 1991.

Smalley RE et al, Fullerenes with metals inside, *J Phys Chem*, 1991.

Smalley RE, Self-assembly of the Fullerenes, *Acc Chem Res* 25:98-105, 1992.

Smith DPE & Quate CF. Atomic point-contact imaging. *Appl Phys Lett* 49(18), 1166-1168,1986.

Smith SA, Watt RC & Hameroff SR, Cellular automata in cytoskeletal lattices. *Physica* 10D:168-174, 1984.

Sneddon SF et al. A new classification of amino acid side chains based on doublet acceptor energy levels. *Biophysical Journal* 53:83-89, 1988.

Soifer D, Factors regulating the presence of microtubules in cells. In: <u>Dynamic Aspects of Microtubule Biology</u>, ed. D Soifer, Ann NY Acad Sci 466:1-7, 1986.

Somjen GC, <u>Neurophysiology, The Essentials</u>. Williams and Wilkins, Baltimore, 1983.

Soureshi R & Wormley D, eds, <u>NSF/EPRI Workshop on Intelligent Control Systems, Final Report</u>, October 1990.

Stebbings H & Hunt C, The nature of the clear zone around microtubules. *Cell Tissues Res* 227: 609-617, 1982.

Stipp D. Researchers find a way to form film of diamond using carbon "buckyballs" *The Wall Street Journal* Wed, Nov. 6, 1991, p. B7.

Stobbs WM Ed. <u>Scanning Tunneling Microscopy. Papers from the 3rd Conference of the Royal Microscopical Society</u>. Blackwell Scientific, Oxford, 1989.

Stroscio J & Eigler D, Atomic and molecular manipulation with the scanning tunneling microscope. *Science* 254:1319-1326, 1991.

Sudharsanan SI & Sundareshan MK, Training of a three-layer recurrent neural network for nonlinear input-output mapping, <u>Proceedings of the 1991 Int Joint Conference on Neural Networks (IJCNN-91)</u>, Seattle, WA, 1991.

Sundareshan MK, Exponential stabilization of large-scale systems: Decentralized and multilevel schemes, <u>IEEE Transactions on Systems, Man, and Cybernetics</u>, Vol. SMC-7, pp. 478-484, 1977a.

Sundareshan MK, Reliability considerations in decentrally controlled multivariable systems. <u>Proc. NEC International Forum on Alternatives for Linear Multivariable Control</u>, Chicago, IL, pp. 310-330, 1977b.

Sundareshan MK, Generation of multilevel control and estimation schemes for large scale systems - A perturbational approach, <u>IEEE Transactions on Systems, Man, and Cybernetics</u>, Vol. SMC-7, pp. 144-152, 1977c.

Sundareshan MK & Acharya RS, Optimal control in cancer therapy by simultaneous consideration of normal and cancer cell proliferation kinetics, *Int J Sys Sci* 15:773-776, 1984.

Sundareshan MK & Fundakowski R, Stability and control of a class compartmental system with application to cell proliferation and cancer therapy, *IEEE Transactions on Automatic Control* AC-31:1022-1032, November 1986.

Suresh B & Seehra M, Temperature dependence of the infrared spectra of C_{60}: orientational transition and freezing, *Chem Phys Lett* 196,6:569-572,1992.

Surridge CD & Burns RG, Phosphatidyinosital inhibits microtubule assembly by binding to microtubule-associated protein 2 at a single, specific, high affinity site. *Biochemistry* 31:6140-6144, 1992.

Sutherland R and Hutchinson JS, Three-dimensional NMR imaging using selective excitation. *J Phys* E11:79-83, 1978.

Szu H & Tate A, Synergism Between Molecular and Neural Computing, in press.

Tabony J & Job D, Gravitational symmetry breaking in microtubular dissipative structures. *Proc Natl Acad Sci* 89:6948-6952, 1992.

Tabony J & Job D, Microtubular dissipative structures in biological auto-organization and pattern formation. *Nanobiology* 1(2):131-147, 1992.

Taniguchi N, On the basic concept of nanotechnology, *Proc Int Conf Prod Eng* Tokyo, Part 2:18-23, 1974.

Tannor DJ & Rice SA, Control of selectivity of chemical reactions via control of wavepacket evolution, *J Chem Phys* 83(10), 1985.

Tannor DJ and Rice SA, Coherent pulse sequence control of product formation in chemical reactions. *Adv Chem Phys* 70(5):441-523, 1987.

Tarnai T, The observed form of coated vesicles and a mathematical covering problem. *Mol Biol* 218:485-488, 1991.

Taylor R, Hare JP, Abdul-Sada A & Kroto H, Isolation, separation and characterisation of the Fullerenes C_{60} and C_{70}: The third form of carbon. *J Chem Soc Chem Commun*, 1423-1425, 1990.

Teresco J, Profiles in technology: Buckminsterfullerenes. *Industry Week* Nov. 4, 1991.

Theurkauf WE & Vallee RB, Extensive cAMP-dependent and cAMP-independent phosphorylation of microtubule associated protein 2. *J Biol Chem* 258:7883-7886, 1983.

Thomson TM. From error-correcting codes through sphere packings to simple groups. The Mathematical Association of America Inc., 1983.

Timasheff SN, Melki R, Carlier MF & Pantaloni D, The geometric control of tubulin assemblies: cold depolymerization of microtubules into double rings. *J Cell Biol* 107:243a, 1988.

Toffoli T, Cellular automata as an alternative to (rather than an approximation of) differential equations in modeling physics. *Physica* 10D:117-127, 1984.

Tolbert SH, Alivisatos AP, Lorenzana HE, Kruger MB and Jeanloz R, Raman studies on C_{60} at high pressures and low temperatures, *Chem Phys Lett* 188,3-4:163-167, 1992.

Toyohashi University of Technology produces catalyst from C_{60} fullerene, *Diamond Deposition: Science and Technology*, March 30, 1992, p. 11-12.

Triggiani R, *SIAM J On Control* 13:462, 1975.

Tsukita S, Tsukita S, Kobayashi T & Matsumoto G, Subaxolemmal cytoskeleton in squid giant axon, II. Morphological identification of microtubules and microfilament-associate domains of axolemma. *J Cel Biol* 102:1710-1725, 1986.

Tuszynski JA, Paul R, Chatterjee R & Sreenivasa SR, Relationship between Fröhlich and Davydov models of biological order. *Phys Rev A* 30(5):2666-2675, 1984.

Varela F & Coutinho A, Second generation immune networks, *Immunology Today* 12:159-166, 1991.

Vassilev P, Kanazirska M & Tien HT, Intermembrane linkage mediated by tubulin. *Biochem Biophys Res Comm* 126:559-565, 1985.

Ventila M, Cantor CR, Shekinski M, A circular dichromism study of microtubule protein. *Biochemistry* 11: 1554-1561, 1972.

Verheyen HF. The icosahedral design of the Great Pyramid, in Fivefold Symmetry, ed I. Hargittai, World Scientific, Singapore, 1992 pp. 333-359.

Vivier E, Morphology, taxonomy and general biology of the genus Paramecium. In: The Biology of Paramecium, ed. WJ Van Wagtendonk, Elsevier, Amsterdam, 1-89, 1985.

Voelker MA, Hameroff SR, Jackson D He, Dereniak EL, McCuskey RS, Schneiker CW, Chvapil TA, Bell LS & Weiss LB, STM imaging of molecular collagen and phospholipid membranes. *J of Microscopy* 152(2):557-566, 1988.

Von Hippel. Molecular designing of materials. *Science* 138(12), 91-108, 1962.

Von Neumann J, Theory of self-reproducing automata. Ed. AW Burks, Urbana, University of Illinois Press, 1966.

Walton D & Kroto H, private communication, 1992.

Wang N & Rasenick MM, Tubulin-G protein interactions involve microtubule polymerization domains. *Biochemistry* 30:10957-10965, 1991.

Wang Y, Caspar J, Cheng L-T, Light and Fullerenes: Photoconductive and nonlinear optical properties. *J Electrochem Soc* 139(4), 1992.

Weaver G. Classifying "aleph zero"- Categorical Theories, Studia Logica, XIVII: 4,327-345, 1988.

Weeks DE & Harter WG, Rotation-vibration spectra of icosahedral molecules. II. Icosahedral symmetry, vibrational eigenfrequencies and normal modes of buckminsterfullerene. *J Chem Phys* 90(9):4744-4771, 1989.

Werbos PJ, Beyond regression: new tools for prediction and analysis in the behavioral sciences. Ph.D. thesis, Harvard University, 1974.

Werbos P, Neurocontrol and fuzzy logic: connections and designs, *Int J Approximate Reasoning*, 1991.

Werbos P, Quantum theory, computing and chaotic solutions, *Proc 2nd Int Conf Fuzzy Logic and Neural Networks,* Iizuka, Japan, July 1992.

Werbos P, The cytoskeleton: why it may be crucial to human learning and neurocontrol, *Nanobiology* 1(1)75-95,1992

Wheatley DN, The Centriole: A Central Enigma of Cell Biology.   Amsterdam, Elsevier, 1982.

Whetten RL, Alvarez MM, Anz SJ, Shriver KE, Beck RD, Diederich FN, Rubin Y, Ettl R, Foote CS, Darmanyan AP & Arbogast JW, *Materials Research Society Symposium Proceedings*, RS Averback, J Bernholc & DL Nelson, Eds, Materials Research Society, Boston, 639, 1991.

Wichterman R, The Biology of Paramecium, New York, Plenum, 1985.

Wilson AC, & Cann RL, The recent African genesis of humans, *Scientific American* 4:68-73, 1992.

Withers JC & Loutfy RO, U.S. Patent, applied 1992.

Wolfram S, Cellular automata as models of complexity. *Nature* 311:419-424, 1984a.

Wolfram S, Universality and complexity in cellular automata. *Physica D* 10:1-35, 1984b.

Wright R, The on/off universe. The information age. *The Sciences*, May/June, 7-9, 1985.

Wu ZC, Jelski AD, & George FT, Vibrational motions of Buckminsterfullerene, *Chem Phys Letters* 137,3:291-294,1987.

Wuensche A, Basins of attraction in disordered networks. In: Artificial Neural Networks, 2, ed. I Aleksander and J Taylor, Elsevier, Amsterdam, 1325-1344, 1992.

Xie Q, Perez-Cordero E and Zchaegoyen L, Electrochemical detection of C_{60}^{6-} and C_{60}^{6-}:enhanced stability of fullerides in solution. *J Am Chem Soc* 114:3978, 1992.

Yoshida Z, Dogane I, Ikehira H, Endo T, Investigation of hydrogenation of C_{60} using theoretical calculations. *Chem Phys Lett* 201,5-6:481-485, 1992.

Zadeh LA. Fuzzy sets, information and control, 8:338-353, 1965.

Zangwill WI, Nonlinear Programming: A Unified Approach, Prentice-Hall, Englewood Cliffs, New Jersey, 1969.

Zheluder IS. Complete symmetry of figures with fivefold symmetry axes in Fivefold Symmetry, ed I. Hargittai, World Scientific, Singapore, 1992 pp. 171-176.

Zhou F, Yan SL, Jehoulet C, Laudi Jr. DA, Guan Z & Bard AJ, Electrochemistry of C_{60} films: Quartz crystal microbalance and mass spectrometric studies. *J Phys Chem* 96:4160, 1992.

Zimmerman JA, Eyler JR, Bach LSBH & McElvany SWJ, Magic number carbon clusters: ionization potentials and selective reactivity. *J Chem Phys* 94:3556-62, 1990.

FIGURE SOURCES

1-1: Hatch, A., Buckminster Fuller: At Home in the Universe, Crown Pub. Inc., New York, 1974. **1-2**: Fuller, B.R., Synergetics: Explorations in the Geometry of Thinking, MacMillan Publishing Company, New York, 1975. **1-3**: ibid 1-2. **1-4**: Adapted from (1) Barth W.E. and Lawton R.G. 1971, (2) Osawa, 1970, (3) Osawa E, the Royal Society Meeting (London) on Post Buckminsterfullerene view of the Chemistry, Physics and Astrophysics of Carbon, October. **1-5**: Adapted from (1) Eaton P.E., 1992 and Mueller R.H., 1972, (2) Buchvar D.A. and Gal'pern E.G., 1973. **1-6**: Adapted from (1) Rohlfing E.A., et al., 1984, (2) Kroto H.W. et al., 1985. **1-7**: Adapted from (1) Day K.L. and Huffman D.R., 1973, (2) Huffman D., 1991, (3) Kräschmer W. et al., 1990. **1-9**: Adapted from Nature (cover page) October 22, 1992. **1-10**: Adapted from Zare R. H., Angular Momentum, p. 351, John Wiley & Sons, 1988. **1-11**: Adapted from Kanadeki T and Kadota K, 1969. **1-12**: Adapted from Science (cover page) December 20, 1991. **1-14**: Adapted from Buseck P.R. et al., 1992. **2-1**: Adapted from Hargittai, 1992. **2-2**: ibid 2-1. **2-3**: ibid 2-1. **2-4**: Adapted from Clark R.T.R., Myth and Symbol in Ancient Egypt, Thames and Hudson, 1959. **2-5**: ibid 2-1. **2-10**: ibid 2-1. **2-19**: ibid 2-1. **3.2-4**: Adapted from (1) Seaito, 1991, (2) Ref. Satpathy, 1986. **3.2-5**: Adapted from Pasqnarello A et al., 1992. **3.3-1**: Adapted from (1) Weeks D.E. and Harter W.G. 1989 (2) Wu Z.C. et al, 1987. **3.3-2**: Adapted from (1) Krätschmer et al, 1990, (2) Saito, S. 1991. **4.2-1**: Haddon R.C. 1988. **4.3-2**: Harter W.G. and Weeks D.E., 1989. **4.3-3**: ibid 4.3-2. **4.3-4**: ibid 4.3-3. **5.1-1**: Adapted from (1) Kanasaki T. and Kadota K, 1969 and (2) Schmid S.L, 1992. **5.1-2**: Adapted from (1) Schmid S.L., 1992 (2) Heuser J, and Kirchhausen, T, 1985. **5.2-1**: Adapted from Dustin, 1984. **5.2-2,3,4**: Ref. ERickson, 1973. **5.3-2** Adapated from: Sloane N.J.W, 1984. **5.3-3**: Hirokawa, 1991. **5.3-6**: Adapted from (1) Bailly and Bornens, 1992 and (2) Alberts et al., 1990. **5.3-7**: Adapted from Dustin, 1984. **5.3-8**: Bornens, 1979. **6-1**: Dryl, 1974. **6-2**: Vivier, 1986. **6-4**: Wichterman, 1985. **6.5**: Cohen and Beisson, 1988. **6-6**: Hirokawa, 1991. **6-11**: Rasmossen et al., 1990. **6-12**: Rasmossen et al., 1990. **6-14**: Rasmossen et al., 1990. **6-16**: Wuensche and Lesser, 1992. **7.3-1**: Adapted from: The Golden Age of Cosmology, Scientific American, 7, pp. 17-20, 1992. **8-8**: Adapted from (1) Iijima, 1980, (2) Dresselhaus M.S., 1992. **8-9**: From (1) Dresselhaus M.S., 1992, (2) Iigima, 1991. **8-10**: Adapted from (1) Curel and Smalley, 1991 (2) Kroto et al, 1985. **9.1-1**: Adapted from Morton et al., 1992. **10-1**: Dubois & Kadish, 1991. **10-2**: Xie & Perez-Cordero, 1992.

GLOSSARY

AFM: [Atomic Force Microscopy] - derivative of STM (scanning tunneling microscopy) in which scanning tip is deflected by mechanical touch between tip and sample. Better than STM for non-conducting materials, can yield atomic resolution. AFM was invented by Binnig in 1986.

ATP: [Adenosine Triphosphate] - biochemical energy source which functions as a phosphate-group donor; supplies 0.2 electron volts per phosphate group for mechanical (muscle, cytoskeletal) movement and metabolic needs.

Battery: One or more electrochemical cells electrically interconnected to form a single unit which has provisions for external connections.

Bias: A voltage potential applied between STM tip and sample to produce desired characteristic.

Bio-Fullerenes: Biological structures with either icosahedron or fivefold symmetry (Golden Mean property). A bio-Fullerene with icosahedron symmetry is clathrin (Golden Mean-ϕ^3); a bio-Fullerene with a fivefold symmetry, as an icosahedron shadow, is the microtubule (Golden Mean ϕ^{-2}).

C_{60} molecule: Fullerene "buckyball" - carbon cluster of 60 atoms with truncated icosahedron symmetry. Theoretically prediced by Osawa (1970), discovered by Kroto/Smalley research team (1985), produced by Huffman/Krätschmer research team (1973-1989) and first imaged with atomic resolution by Koruga/Hameroff research team (Jovana Simic-Krstic, Mirko Trifunovic, 1992).

C_{60}-NMT: [C_{60}-New Moon Technology] - A concept of C_{60} nanotechnology which involves production of optical and/or electro-optical devices which can be compatible with, and interface to, human information processing.

C_{116}: Dimer of C_{60} molecule with four bonds between monomers. Product of reaction series: $2C_{60} - 2C_2 \rightarrow 2C_{58} \rightarrow C_{116}$ (strong dimer). Initially created and proposed by D. Koruga using scanning tunneling engineering (STE).

Charge density: Distribution of molecular electronic charge density calculated as sum of molecular orbital densities, each the square of the orbital wave function.

Clathrin: 180,000 molecular weight protein with truncated icosahedral symmetry identical to that of the C_{60} molecule. The diameter of clathrins range from 30 to 60 nm.

Crystal symmetry: A class of the point symmetry groups known as the Bravais lattice which provides 7 classes of the 32 possible point groups

Curie symmetry: Originally proposed by Marie and Pierre Curie in 1894, Curie symmetry represents a class of point symmetry groups which includes the infinite symmetry axes and is based on the physical properties of crystals. There are seven Curie symmetry groups which represent crystals in a homogeneous, continuous medium. Curie symmetry relates the structure and physical properties of microtubules to bio-Fullerenes.

Dimer: Two equal or similar subunits which, when joined together, produce a structure having new properties.

Degeneracy: The number of identical energy states. As molecular symmetry increases, the level of degeneracy also increases.

DNA: [Deoxyribonucleic Acid] - A double-helical polynucleotide having a base sequence encoding genetic information. DNA (and other nucleic acids-i.e. RNA) exhibit right optical activity. The current scientific dogma is that genetic information flows from DNA to RNA to protein. DNA structure was identified by Watson and Crick in 1953.

Electrode: Conducting body within a cell on which electrochemical reactions occur. It normally consists of an active material and those structures necessary to collect the accumulated charges and which can support the active material.

Electron: An elementary particle possessing a negative charge (1.602×10^{-19} C) and a rest mass of 9.109×10^{-31} kg.

Energy surface: A multidimensional plot of the potential energy of a molecular system as a function of all variables in the system.

Five-fold symmetry: Property of a subject (molecule, process, etc.) for which symmetry operations produce new orientations equivalent to the original orientation five times.

Fullerenes: General name for three dimensional carbon clusters which contain more than 24 carbon atoms with 12 pentagons and a variable number of hexagons.

Fullerite: Solid   crystalline   form   of   fullerene.   (Term   proposed   by Huffman/Krätschmer, 1990).

Golden Mean: Value of two inversion Fibonacci series which generates $\phi = (\sqrt{5} + 1)/2$ and $(1/\phi = (\sqrt{5}-1)/2$ with each property as $\phi = 1 + 1/\phi$ (also notation τ).

Holopent: Synergy of information as a space-time pattern based on five-fold symmetry. (Greek: *holo* - whole, *pent* - five; term proposed by Koruga, 1991).

HOMO: [Highest Occupied Molecular Orbital] - highest occupied electron state of configuration of atomic orbitals in a molecule.   Used as a reference point for describing the position of a molecular orbital as being at or below this level.

Hückel: A simple and approximate method for performing semi-empirical quantum mechanics calculations.

HWK: [Harter-Weeks-Koruga] - A theoretical-experimental model of C_{60} based on tensor-operator methods and STM images of the C_{60} energy surface.

Hydrogenation: Doping a molecule (i.e. C_{60}) with hydrogen from outside and/or inside of the molecule.

Icosahedron: One of the five Platonic solids which has 20 faces (equilateral triangles), 12 vertices and 30 edges.   Schoenfies symbols are I - for an icosahedron with order 60 symmetrical operations and I_h - for an icosahedron with order 120 symmetrical operations including reflections and inversions. Truncation of vertices produces C_{60} geometry.

Jahn-Teller Effect: The energy in nonlinear molecules is minimized when distortions occur to remove either spin or orbital degeneracy or both. An exception occurs for those systems which have a spin with two-fold degeneracy of the Kremers type.

Kugel Symmetry: The symmetry group which represents spherical harmonics and corresponds to the symmetry of a sphere with rotation-inversion properties.

LUMO: [Lowest Unoccupied Molecular Orbital] - Lowest unoccupied electron state of configuration of atomic orbitals in a molecule. Used as a reference point for describing the position of any molecular orbital as being at or above this level.

Microtubule: Major component of cell cytoskeleton. Hollow cylindrical structure usually composed of 13 protofilaments, which are comprised of α and β tubulin subunits arranged according to the Golden Mean law.   Microtubules organize intracellular dynamics and information.

Mind Synergy: A manner of thinking by way of integration of human consciousness and unconsciousness. Information processing which solves a problem in an unexpected way and/or results in an unanticipated discovery.

Molecular Orbital: A configuration of atomic orbitals. Describes the electron state of molecules.

Nanobiology: Biochemical/biophysical discipline which includes the study of mass, energy, information, organization and control in the nanoscale range [nano $= 10^{-9}$; nanometer (nm) $= 10^{-9}$ m; nanosecond (ns) $= 10^{-9}$ sec].

Nanotechnology: A new scientific and engineering paradigm which involves the manipulation of atoms and molecules in order to produce new, improved materials and/or molecular devices.

Negative Electrode: An electrode which maintains a potential lower than that of a second electrode during normal cell operation.

NSN: [Njegoš Scenario Number-Φ!]-A number which is so large that only it can adequately represent itself (N > 0) and also so small that only it can adequately represent itself (N < 0). Njegoš used this number in his poem, "The Ray of the Microcosm" (Luča Mikrokozma) which was written on the Golden Mean. In his poem, Njegoš constructed the *scenario* ($\infty^{\perp}$) which he develops through permutations based on the Golden Mean. For example, although the Fibonacci series is normally written as: 0,1,1,2,3,5 ..., according to NSN it should be as follows: 0,0!, 1,2,3,5 ... The relations, $R^N \cdot S_w^{(1-N)} = N(0)$ when written according to the sense of the Golden Mean is as: $\phi^N \cdot \phi^{(1-N)} = \phi^{0!}$ repectively. The entity $\phi^{0!} = \Phi$! is the Njegoš Scenario Number. This number [Φ! $\equiv$ N(0)] is the source of *creation* (as described in Njegoš' "***The Ray of the Microcosm***") and is the crown of true God, o'er all times and o'er all eternity as he crowned himself. In this poem, Njegoš was saved and managed to overcome a religious way of thinking in the sense of Hegel's *aufgehoben*.

Pi (π)-Electrons: Value of the probability density of a molecular orbital when the probability density is not symmetrical about the line at the centers of the nuclei.

Positive Electrode: An electrode which maintains a potential higher than that of a second electrode during normal cell operation.

Real World: Space-time structure which exists independent of human beings. The human world is a four dimensional space-time structure (3-dimensional, time and light) with mass, energy and information as its main elements.

Red Shift: A displacement of lines in the spectra of certain celestial objects toward longer wavelengths (i.e. toward the red end of the visible spectrum).

Self-assembly: Spontaneous assembly of parts of an ordered system based on the properties of its parts and the whole (synergy of information).

Serendipity: Unexpected, accidental discoveries in science, technology, engineering or life in general.

Shadow: The result of an interception of light by an opaque obstacle. As light rays travel in straight lines, light from the source cannot be detected behind the obstacle.

Shadow Icosahedron: Complementary pattern of an icosahedron based on symmetry inversion of a unit sphere and the Golden mean. The Real (3-D) icosahedron has ϕ^3 Golden mean notation, while the Shadow icosahedron has ϕ^{-2} notation.

Shadow World: Complementary world of any $N > 1$ dimensional world based on symmetry inversion of a unit sphere.

Synergy: The behavior of whole systems not predicted by behaviors or characteristics of any of the system's components when analyzed individually.

Spectrum: A range of electromagnetic radiation emitted or absorbed by a substance under particular circumstances.

Spectroscopy: The study of the interaction of electromagnetic radiation (light, radio waves, x-rays, etc) with matter.

STE: [Scanning Tunnelling Engineering] - A technique based on STM which allows for the manipulation of atoms and molecules to construct or alter a pattern or object through operator design.

STM: [Scanning Tunnelling Microscopy] - A microscopic technique based on electron tunnelling principles. Imaging of samples is possible with atomic resolution. Can be used for imaging conducting and semi-conducting materials.

Symmetry: The basic feature of space-time structures (mass, information and organization) and processes (energy, information and control).

Symmetry Element: A point, line or plane upon which a symmetry operation is based.

Symmetry Operation: An operation which moves a subject (i.e. molecule) under consideration (mass, process, information...) into a new orientation which is equivalent to its original one.

3H Technology: Hand-Head-Heart technology which is evolving from computer technology. Will allow humans to communicate with computers by *hand* (keyboard), *head* (indirect interfacing by voice and/or direct interfacing by electromagnetic devices) and *heart* (direct interfacing through a field which involves the coding and decoding of the human mind and/or emotions).

$2RC_{60}$: Dimer of C_{60} molecule with one bond between monomers with dissociation process: $RC_{60}C_{60}R \rightleftharpoons 2RC_{60}$ (weak dimer), where R can be H,F or CH_3). Proposed by the Morton research group, 1992.

INDEX